The Telecom Revolution in India

The Telecom Revolution in India

Technology, Regulation, and Policy

Varadharajan Sridhar

OXFORD

UNIVERSITY PRESS

OXFORD
UNIVERSITY PRESS

Oxford University Press is a department of the University of Oxford.
It furthers the University's objective of excellence in research, scholarship,
and education by publishing worldwide. Oxford is a registered trademark of
Oxford University Press in the UK and in certain other countries

Published in India
by Oxford University Press
YMCA Library Building, 1 Jai Singh Road, New Delhi 110 001, India

ISBN-13: 978-0-19-807553-0
ISBN-10: 0-19-807553-7

Typeset in Adobe Caslon Pro 10/12.5
by Sai Graphic Design, New Delhi 110 055
Printed in India at G.H. Prints Pvt. Ltd., New Delhi 110 020

To
my wife Kala, daughter Vindhya, and my parents

Contents

Tables, Figures, and Boxes

Tables

Figures

Boxes

Preface

In India, telecommunications has been one of the star performers, post-liberalization. Telecommunications has changed the way people live their daily lives, whether rich or poor, young or old, male or female, urban or rural, literate or illiterate, professional or novice. Though the mobile phone has been the critical technology intervention in this transformation, regulatory reforms, competition, and technology leapfrogging have enabled many other communication sub-sectors to grow as well. My objective was to compile technology, market, regulation, and policy dimensions that impact each telecom sub-sector and provide a comprehensive view to the reader. I have also tried to highlight critical factors that are unique to each sub-sector and show how the triad of technology, regulation, and policy can help optimize these factors for enabling growth of the sub-sector.

The period of my relocation to India from the USA in 1999 was apt for research in telecom. The revenue sharing regime was introduced for cellular mobile services and the New Telecom Policy 1999 was announced by the Government of India. From then onwards it is has been a dream run for telecom in the country, though with regulatory and policy fluctuations, and associated development of appropriate strategies by the concerned entities. My first article about the Indian telecom industry was on international accounting rate that appeared in the *Economic Times* on 8 September 2000 when I was on the faculty of the Indian Institute of Management Lucknow (IIML), India. As predicted in the article, the international settlement regime was slowly dismantled and in 2002 India got out, thanks to the introduction of competition in the international long-distance market. From then on, I worked with many of my colleagues in various institutions and the industry on issues ranging from technology, strategy, markets, theories, regulation, and policy issues in telecom. Being a member of the Committee for Allocation of Access (GSM/CDMA) Spectrum and Pricing of the Department of Telecommunications, Ministry of Communications and IT, Government of India during 2008–9, the assignment put me in touch directly with the leading academicians and policymakers in the country. This book is a compendium of more than 125 articles that I have written in various newspapers, magazines, refereed journal publications, and book chapters on different themes, each suitably revised and updated.

I always had the luxury of teaching elective courses in telecom in various institutions including Ohio University and American University in the USA; University of Auckland in New Zealand; S.P. Jain Centre of Management in Singapore and Dubai; IIML, Indian Institute of Management Bangalore, and Management Development Institute, Gurgaon, in India. The interactions with the students always helped me to hone my ideas whether for an opinion article in newspapers or an indepth research paper. The book captures some of these interactions and valuable feedback from my students on burning issues in telecom. The two summer stints I did at the Department of Communications and Networking at Aalto University, Finland helped me accumulate knowledge about the European market, especially that of Finland which has been the pioneer for most of the breakthrough in technology adoption and regulatory interventions in the wireless space.

After my long stint in academics, when I moved to the industry, Sasken gave me the opportunity to understand the entire supply chain ecosystem in telecom, part of which are covered in chapters 1, 9, and 10 in the book.

An important aspect of telecom that is always exciting is that there are innovations each minute and the industry evolves and morphs continuously. Since I finished my manuscript, sea changes have occurred in the mobile Internet and platform space. It is very difficult to assimilate all the changes as they are moving targets. I hope that I have captured as many as possible to make this an interesting, informative, and insightful experience for the readers.

We will have more Internet, larger numbers of users, more mobile access, more speed, more things online and more appliances we can control over the Internet.

—Vinton Cerf, Father of the Internet

Acknowledgements

Though I have been writing articles on telecom in business newspapers, magazines, and refereed journals over the last decade, this concerted effort of putting it all together in a book would not have been possible without inspiration from my wife Kala. Thanks to her constant persuasion and encouragement, I was able to complete different milestones of this long and arduous project. The theoretical basis of my research in telecommunications was set during my PhD work at the Tippie College of Business at the University of Iowa, USA. Thanks to my thesis supervisor June Park, who spent countless hours correcting the mathematical models and my thesis manuscript, I developed deep research insights. The book also covers basic technologies in telecommunications, the expertise for which I owe to the McClure School of Communication Systems Management, Ohio University, Athens, Ohio, and Kogod School of Business, American University, Washington DC, USA, where I taught courses in telecom technology and management.

I have been fortunate to co-author many articles with my colleagues—foremost amongst them is Rohit Prasad at the Management Development Institute (MDI), Gurgaon, India, with whom I worked on spectrum management issues—which has resulted in many publications some of which are covered in Chapter 4 of the book; G. Venkatesh, CTO of Sasken Communication Technologies, India, with whom I co-teach Telecom Strategy course and with whom I wrote a number of articles on mobile applications, net neutrality, and two-sided markets; S.R. Raja, Associate Vice President at Sasken Communication Technologies with whom I collaborated on a number of articles and projects on mobile value-added service; and my wife Kala Seetharam Sridhar, Senior

Research Fellow at Public Affairs Centre, Bangalore, India, with whom I wrote a number of articles, refereed papers, and book chapters related to telecom and economic development. There are also others including Deepak Maheshwari of Microsoft; G Krishnakumar of Teleca; Parag Kar of Qualcomm; Pramod Pagare of Persistent Systems; Siddharth Gaikwad, Subhash P, and Swami Krishnan of Sasken; and Thomas Casey at Aalto University, Finland, who helped me shape research issues on various thematic areas.

Piyush Jain, my first PhD student at the Indian Institute of Management Lucknow (IIML), India, did an excellent job on the analysis of basic telecom market using system dynamics, a part of which is presented in Chapter 2. I also benefitted from the interactions that I had with my students of telecom courses of the postgraduate programme, and executive management programmes at IIML, MDI, S.P. Jain Centre of Management, and Indian Institute of Management Bangalore, India.

I am thankful to Pritam Singh, former Director of IIML, and MDI for encouraging me to write topical articles in business newspapers and magazines. Interactions with Professor Heikki Hämmäinen of Aalto University, Finland, during my visit as Nokia Visiting Research Fellow helped me to get an international perspective. For many of the articles cited in the book, I am thankful to the editors of *Economic Times*, *Hindu Business Line*, *Business Standard*, and *Financial Express* who went through them patiently and gave valuable feedback. Ibrahim Ahmad, Group Editor of Cyber Media, and Shyamanuja Das, Editor, DataQuest, gave excellent opportunities to my students at IIML for publishing their work in *Voice and Data*, some of which are cited in the book.

My sincere thanks to the anonymous reviewer whose comments helped me introduce Chapter 9 on telecom manufacturing and incorporate the political economy perspective throughout the book. I thank the editorial team at Oxford University Press, India, which worked very closely with me during this project and diligently coordinated the various phases to bring out the book on time.

Finally, my sincere thanks to Rajiv Mody, Chairman and CEO of Sasken Communication Technologies, who gave me the freedom to pursue my writing which is often rare in a corporate environment.

1

Network Economics in Telecom

Some goods and services generate more value when more users consume the same goods and services. The consumers using these products constitute networks in which the utility derived from consumption of these goods or services increases as additional consumers purchase or use the same goods and services. A market characterized by such properties is called a network market in which there exist positive consumption externalities termed 'network externalities'. Examples of products exhibiting network externalities include fax machines, credit card networks, telephone services, broadcast industry services, computer hardware and software. Hence, network externality is defined as the increasing utility that a user derives from consumption of a product or a service as the number of other users who consume the same product or service increases (Katz and Shapiro, 1985).

For example, the telephone was of little value to the first individual to have one. However, with each additional telephone adopter, this innovation became more valuable to all of its users. Thus for users considering adoption of such technologies, the potential return they can realize depends on the number of existing users. This relationship is especially strong for various telephone and computer networks.

Network Effects and Interconnection

In the early 1900s, in most countries including the United States (US) and Finland there were several hundreds of telephone companies providing telecom service. For example, in Finland, there were more than

800 telephone companies providing service during the period 1935–50 (Steinbock, 2001). These companies often refused to interconnect with one another, and each had its own set of subscribers. Few subscribers, of course, wanted to buy several telephones and pay subscription charges to several telephone companies simply to make sure that they could reach anyone else they wished to call. Such arrangements are quite wasteful in that they misallocate society's scarce resources away from more productive uses. A single subscription enabled by interconnection of various networks allows for exchange of traffic at reasonable rates bringing value to the subscriber compared to multiple subscriptions..

In the absence of regulatory interconnection obligation, virtually every telephone market in the early twentieth century in the US and rest of the world reached a 'tipping point', in which the largest network—the one with the greatest number of subscribers—was perceived as the single network that everyone had to join, and the rest withered away. The potential for certain industries to slide into monopoly in this manner illustrates the economic implications of network effects (Nuechterlein and Weiser, 2005).

The applicability of network externalities is very much evident in mobile services in India. After the first cellular licenses were issued in 1995, the subscriber base shot up from nothing to more than 300,000 in just one year. The cellular subscribers could connect and talk to the then existing 12 million landline users and therefore critical mass was achieved almost instantaneously resulting in a good growth in the cellular service adoption in the country. If we had a policy imposing the restriction that cellular subscribers could only talk to other cellular subscribers, the industry would have died instantly due to the absence of critical mass (Sridhar, 2004).

Complementarity and Indirect Network Effects

Many products have little value when used in isolation. They are required to be used with other products at the same time. Computers as standalone machines are of little use unless installed with appropriate software. Computers and software complement each other and thus create network externality effects. Those who have computers need to necessarily buy appropriate software such as operating systems, thus indirectly increasing the externality effect. Such products which are strongly complementary also exhibit positive network effects. Detailed exposition of network effects is illustrated in Yang (1997). There are various network components such as access networks, back-haul networks, submarine cable networks, and switches and routers that complement each other to provide services such as national long distance and international long distance services. Multimedia content and aggregation of such content complements the network components in the provisioning of Internet-based multimedia services. These complementarities enable indirect network externality effects (Economides, 1996).

An innovative Closed User Group (CUG) service called 'Friends and Family' (F and F), being offered by some of the mobile operators, allows subscribers the option to choose one or few numbers, calls to which are made either totally free or are priced at discounted rates. However, if you subscribe to this plan, you can also call numbers not listed in your F and F plan. This flexibility makes F and F a complementary product. If there are two or more complementary goods/services, then adoption of one will increase the adoption of the other and vice versa leading to near exponential growth. Presently, most of the F and F schemes are restricted to numbers within the network of the operator. The Telecom Regulatory Authority of India (TRAI) in a consultation paper advised the operators to extend the choice of numbers to any network to improve network externality effects. Another case of network externalities in operation is prepaid vis-à-vis post-paid schemes adopted by mobile operators. Most of the mobile operators did active selling of their prepaid service initially to acquire subscribers. Once the critical mass was attained, service providers such as Bharti started migrating their prepaid Airtel subscribers to more revenue-earning and less attrition-oriented post-paid schemes.

Some of the complementary goods/services are immediately combinable because of their inherent properties. However, for many complex products, actual complementarity can be achieved only through the adherence to specific technical *compatibility* standards. It is the compatibility that makes complementarity actual (Economides, 1996). In telecommunications this has resulted in the birth of standards such as Global Systems for Mobile (GSM), Wideband Code Division Multiple Access (WCDMA) in mobile services; Data Over Cable Service Interface Specification (DOCSIS) for broadband over cable; and Transmission Control Protocol (TCP)/Internet Protocol (IP) in the case of the Internet.

The standards have roots in industry alliances such as the Symbian Foundation and Linux for Mobile in handset software development; GSM Association and the Code Division Multiple Access (CDMA) Development Group in case of networks and so on. Economides (1996) argues that a firm benefits from a move to compatibility if (i) the marginal externalities are strong; and (ii) it does not increase competition to a significant degree by its action. On the other hand, the coalition benefits from a firm joining its 'standard' if (i) the marginal externality is strong; (ii) the firm that joins the coalition is large; (iii) competition does not increase significantly as a result of the firm joining the coalition. Since the second and third criteria may create incentives that are in conflict, this will help define the equilibrium coalition structure. Katz and Shapiro (1985) show that if the costs of achieving compatibility are lower for all firms than the increase in profits because of compatibility, then the industry move toward compatibility is socially beneficial. Sridhar and Raja (10 September 2008) argue how network externalities and complementarity were reasons for the formation of the Symbian Foundation in the mobile handset software space (see Box 1.1). Sridhar and Venkatesh (1 March 2010) discuss competition in mobile platforms (Symbian, Android, iOS) and argue that building direct and indirect network effects are essential for the success of a platform.

Box 1.1   Network Effects for the Success of Mobile Application Platforms

Microsoft announced in 2008, Windows Mobile 7 Series platform along with its strategic partners HTC, LG and Orange; Alcatel-Lucent, the telecom gear-maker, made an unlikely move into the mobile applications development platform based on cloud computing referred to as 'sandbox in the sky'. Similarly, there are a host of application development communities such as the 'Open Handset Alliance' which produced the Android mobile operating system; development workgroups such as the 'Joint Innovation Lab' formed by China Mobile, Verizon and others; the 'App Store' of Apple; Nokia's 'Ovi' platform; 'Flypp' of Infosys; and the list goes on and on. However, the irony is that in this world of mobile phone technologies, incompatibility still persists. For example, an application developed for Apple's iPhone does not work without specific modifications on Google's Android-based Nexus One, nor on Nokia's Symbian-based phones. Hence the moot question is where will all this lead to and who will succeed in this cacophonic marketplace?

These mobile technologies and application frameworks conceptualized and produced by a group of players can be termed as industry platforms that have two important features. The platforms provide a way for different firms/ players that are part of the associated group to contribute and 'complement' each other. This complementarity of components is important for the survival of the platform. The smartphone is of little value if there are no applications for the user to experience. Complementarity of components developed for the platform breeds 'network effects'. There are positive feedback loops that can grow at geometrically increasing rates as adoption of the platform and the complements rise. This dynamic, driven by network effects, encourages more users to adopt the platform, and more complementers to enter the ecosystem to build applications, almost ad infinitum.

Traditionally, the mobile ecosystem can be conceptualized as a two-sided market where on one side the application/platform developers make contracts with the mobile service provider for provisioning their content/applications while on the other side the mobile operator who has control over the 'last mile of access' trades them to the end subscribers. However, recently, the platform developers, such as Apple, have been able to nurture such a strong ecosystem between the developers and users with large network effects that it poses a threat to the role of the operator—the middleman. By enabling the subscribers to download applications directly from the Appstore, and by retailing Nexus One through its web store, Apple and Google respectively are breaking the 'walled garden' of the operators.

Moreover, the innovative platforms and the associated applications drive up data traffic on the network of the operators. With a small proportion of bandwidth-hungry users hogging up the majority of the network capacity, the operators are ill-equipped to provide guaranteed quality of service to their remaining normal users. This is precisely the reason why operators have also joined the bandwagon of creating platforms to regain control over what gets used on their networks for better network planning. About twenty operators including Bharti Airtel, China Mobile, and Vodafone have formed an open system for the development and distribution of mobile applications referred to as the 'Wholesale Applications Community'.

The renowned economist Nicholas Economides noted that both the firm joining the coalition and the coalition benefit if (i) marginal externalities are strong and (ii) competition does not increase significantly as a result of the firm joining the coalition. By sharing the cost of development of a unifying mobile platform among its members, coalitions such as the above platform communities intend to commoditize the mobile applications and services. The communities practice the 'develop and share' philosophy, contributing to strong marginal externalities.

We are still in the early stage of witnessing the development of platforms and community ecosystems in the mobile handset space. Taking a clue from the quotes of Cusumano at the MIT Sloan School of Management, who wins and loses in this competitive game of mobile handset platforms, is not a simple matter of who has the best technology or who is the 'first mover'. It could be the one who has the best platform/community strategy and has the best ecosystem consisting of both the developers and users, to back it up.

Economies of Scale, Density, and Scope

The 'network effect' and 'scale economies' are closely related to each other. Each describes a characteristic of markets in which, all else held constant, increasing the scale of a firm's operations improves the ratio of (i) the value of the firm's services to each customer, and thus the revenues the firm can obtain from that customer, to (ii) the per-customer cost to the firm for providing those services (Nuechterlein and Weiser, 2005). Network effects improve this ratio by increasing the value of the service to each customer, whereas the scale economies improve it by decreasing the per-customer cost of providing that service. In the absence of regulation, each result would play a powerful role in favouring larger-scale telecom firms over their smaller rivals.

Interconnection obligations significantly lower the entry barriers posed by the combination of network effects and scale economies. The new entrants do not have to build ubiquitous networks while competing with incumbents. Interconnection obligations reduce any advantage the incumbents have from network effects. However, the scale economies enjoyed by the incumbents over new entrants still remain.

The commonly held view is that competition is the most effective market structure to ensure low prices and high quality. However, in industries such as fixed line telecom services and electricity distribution, economies of scale and scope are large enough to warrant low levels of competition, even monopolies, to minimize unit costs. Telecom carriers face huge initial costs, including, for example, laying down copper lines from the Central Office to each subscriber location in case of basic fixed line services, constructing cell sites and Base Transceiver Stations (BTSs) in case of mobile services, and laying optic fibre cables to interconnect their access networks to backbone networks. These costs are both fixed, in that the operator must incur them upfront before it can provide any volume of service, and sunk, in that, once incurred, the investment cannot be put to some other use. In contrast, the marginal cost of providing services to each additional customer, once the network is operational is often less. Given the enormous fixed costs and negligible marginal costs, the carrier's long-run average costs within the defined geographical area may well decline with every increase in the size of the network. In other words, it is often cheaper for an operator to provide services to the one-millionth customer than to the one-thousandth customer (Nuechterlein and Weiser, 2005).

The presence of economies of scale poses a ticklish question for the votaries of competition. Allowing a few firms to dominate a market would lead to greater efficiency in production, but at the risk of increased mark-ups over marginal cost on account of market power. Prasad and Sridhar (2008) established the presence of economies of scale in the Indian mobile industry and then estimated the number of operators for the optimal functioning of the market. They found that the level of competition and the number of mobile operators in each service area were greater than that required for attaining scale economies.

Closely related to economies of scale are economies of density. Suppose that the cost of digging and laying down transmission and local access loop to a defined geographical region from the nearby exchange is about Rs 100,000. If the area happens to house an apartment complex with about 1,000 residences, the sunk cost incurred by a monopoly incumbent per residential customer would be Rs 100. If there are two operators providing fixed line connection to this complex, the new entrant would also incur the same Rs 100,000 in providing the local access loop with the result that the cost per residential customer would increase to Rs 200. However, if the area selected by the operators for providing the service happens to have individual houses cumulating to about 100, the cost per residential customer would turn out to be Rs 1,000 and Rs 2,000 respectively in monopoly and duopoly market conditions. This example illustrates that economies of scale within a geographical location (that is, more subscribers within a defined area) yields lower per customer cost. This phenomenon is often referred to as 'economies of density'. Economies of density are cited as one of the reasons for basic wire-line services not reaching the rural areas despite introducing competition. Sridhar and Malik (2007) illustrate how new entrants are forced to deploy wire-line service in dense urban areas of the country to minimize unit costs.

The third important factor in telecommunications is 'economies of scope' wherein firms expand their product/service offerings that complement each other. Whereas economies of scale primarily refer to efficiencies associated with supply-side changes, such as increasing or decreasing the scale of production of a single product type, economies of scope refer to efficiencies primarily associated with demand-side changes, such as increasing or decreasing the scope of marketing and distribution, of different types of products. Economies of scope are one of the main reasons for marketing strategies such as product bundling, product lining, and family branding. Often, as the number of products promoted is increased and different media (radio, television, Internet) are used for promotion, more people can be reached with each rupee spent for marketing and distribution expenses. This is one example of economies of scope. By bundling telephony with Internet broadband, the basic wire-line service providers attain economies of scope, thus reducing cost per unit of products sold. While in the single-output case, economies of scale are a sufficient condition for the verification of a natural monopoly, in the multi-output case, they are neither necessary nor sufficient. Economies of scope are, however, a necessary condition. As a matter of simplification, it is generally accepted that, should economies of scale and of scope both apply, then a natural monopoly exists, which has been perceived to exist in the basic wire-line telecommunication services. In such a scenario, the government traditionally addressed such a market by awarding a monopoly to a single firm and heavily regulating it so that the monopoly does not abuse its market power.

Network Externality Models

Sarnoff was one of the first to come up with a linear model of the network externality effect. Famously known as Sarnoff's law, it states that the value of the network is proportional to the number of entities connected to the network. This has applicability in the broadcast network as the value of the network is in just connecting different households to receive broadcast. Since there is no connectivity between peer households, each household gets a unitary value by connecting to the network, thus making the value of the network to N, the number connected to the network.

Bob Metcalfe, inventor of the Ethernet, was one of the first to point out that the total value of a communications network grows with the square of the number of devices or people it connects (that is, Value N^2). This scaling law, along with Moore's Law invented by Gorden Moore of Intel, is widely credited as the stimulus that has driven the stunning growth of Internet connectivity. Since Metcalfe's law implies that value grows faster than does the (linear) number of a network's access points, merely interconnecting two independent networks creates value that substantially exceeds the original value of the unconnected networks. If there are N nodes in a network, the number of all possible two-way pair-wise communications is $N \times (N-1) = N^2 - 1$. It is to be noted that here all network nodes communicate with each other (that is, connecting peers) and not just with the central node as in the case of the broadcast network. Assuming that each potential connection is worth as much as any other, the value to each user depends on the total size of the network, and the total value of potential connectivity scales much faster than the size of the network, proportional to N^2. Thus the growth of Internet connectivity, and the openness of the Internet, is driven by an inexorable economic logic.

However, David Reed espoused that in networks like the Internet, Group-forming Networks (GFNs) are an important additional kind of network capability. A GFN has functionality that directly enables and supports affiliations (such as interest groups, clubs, meetings, communities) among subsets of its customers. Group tools and technologies (also called community tools) such as user-defined mailing lists, chat rooms, discussion groups, buddy lists, team rooms, trading rooms, user groups, market makers, and auction hosts, all have a common theme—they allow small or large groups of network users to coalesce and to organize their communications around a common interest, issue, or goal. Sadly, the traditional telephone and broadcast/cable network frameworks provide no support for groups. Reed concluded that the GFNs create a new kind of connectivity value that scales exponentially with N. Briefly, the number of non-trivial subsets that can be formed from a set of N members is $N^2 - N - 1$, which grows as N^2. Thus, a network that supports easy group communication has a potential number of groups that can form that grows exponentially with N (for details refer to Reed, 1999).

However, there have been a number of arguments against Metcalf's law and Reed's law. Many contend that since all peer-to-peer connections or group-to-group connections are not of the same value, the value of the network is not $O(N^2)$ or $O(2^N)$, but only $O(NlogN)$.

When the population of potential subscribers is finite, the resulting pattern, which has $O(NlogN)$ growth tends to be S-shaped (Gurbaxani, 1990). The S-shaped curve has three distinct phases. In the initial stages, growth is less as subscribers are not fully informed and aware about the utility they derive by joining the system. However, as the number of subscribers attains a 'critical mass', the number of subscribers in the system is large enough for the growth process to become self-sustaining. Growth increases near exponentially in the second phase. The last stage indicates stability in growth as the saturation level of subscribers is reached. Such an S-curve is illustrated in Figure 1.1. Box 1.2 illustrates the case of Iridium satellite service that failed to reach critical mass and hence had its natural death (WSJ, 18 August 1999).

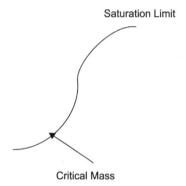

Figure 1.1   The S-curve Model of Growth

Box 1.2   The Iridium Case

Robert Galvin, the legendary son of Motorola's founder and who was then its Chairman, gave the green signal for a project that could be the cornerstone of communications industry, in the 1980s. The ambitious engineers at Motorola's headquarters at Schaumburg, Illinois toiled on the architecture of a system of sixty-six Low Earth Orbit (LEO) satellites placed at a height of about 781 kilometres above the earth's surface, arranged in a necklace configuration covering almost the entire globe, including the north and south poles. It was a technology marvel as the system had sophisticated on-board inter-satellite communication and switching capabilities to switch calls from one satellite to another as the satellite hover over one's head for precisely 59 minutes. The system was targeted as a possible substitute for cellular mobile communications. The total cost of the project was about US$ 5 billion.

With much fanfare, Iridium LLC with support funding from Chinese and Russian governments to the tune of about US$ 2 billion, launched the service on 1 November 1998. However, the satellite service failed to catch the attention of consumers. Nine months after the launch, it had only 20,000 customers. In November 1999, Iridium filed for bankruptcy, much to the chagrin of all technology enthusiasts. It is a classic case of a technology, the adoption of which never reached critical mass and hence was not able to sustain itself. The blunders in the management of satellite communication services of Iridium caused its failure:

Bulky Handset

Since the handset had to communicate with the satellite, the power and signal quality requirement demanded a larger battery pack and a bigger antenna. The clunky handset which weighed more than a pound did not go well with consumers who were already familiar with small packet sized cell phones. Edward Staiano, the CEO of Iridium, just before the company filed for bankruptcy, stormed into a meeting with a large suitcase containing the phone and the accessories and flared at the engineers 'Do you expect a business traveler to carry all these stuff?'. However it was all too late! Moreover, the phone required dexterity in handling. The users had to hold the handset and position themselves so that there is no blocking between the handset antenna and the orbiting satellites. Some of the potential investors who sat through Iridium's presentation were appalled by the forced user behaviour and thought it would never take off.

Expensive and unreliable service

The large satellite handset cost about US$ 7,000 and call costs were as high as US$ 7 per minute. Compare this to the cellular mobile services, through which one could make calls for as low as 10 cents per minute using a US$ 200 smaller handset. Moreover, in a satellite call, there was no distinction between a local call to someone next to you versus to someone in a Siberian oil platform. The calls took almost the same route compared to a cellular mobile call which had to go through many interconnected networks to reach the destination. When a call to Siberia was very much valuable using a satellite phone, the utility a user derived by using a satellite phone to call his friend in the same city was marginal. Though Motorola thought that satellite phone would substitute for cellular mobile services, customers thought otherwise.

The launch was rushed and premature. Handsets were not available in required quantities and technical glitches provided inadequate quality of service to the frustration of the first time users. Compared to the Iridium service, cellular mobile service based on the ubiquitous Global Systems for Mobile was robust and less prone to failure.

Insufficient marketing and distribution

The satellite service was sold through franchisees in different countries. The franchisees also maintained the ground stations (about fifteen of them) for interconnecting the Iridium system with the networks in their respective countries. Though Iridium's global ad campaign appeared in publications all around the world, and, brought in more than a million sales queries, the franchisees were ill-equipped to deal with them. With no marketing channels and a precious few sales people in place, most global partners were unable to follow-up on the inquiries. They did not even have enough supply of handsets to sell to their first-in-line customers. Iridium and Motorola became so obsessed with the technical grandeur that they made fatal marketing mistakes.

The satellite system that was supposed to revolutionize communications by allowing phone calls at any time to anywhere failed miserably. The brassy US$ 100 million international campaign fizzled out and almost all the top executives including Staiano were fired.

Diffusion of Innovation and the Adoption Pattern

The network externality effect and diffusion and adoption of communication technology are closely interlinked. The nature of diffusion and adoption of a new communication technology can be viewed from the perspective of the innovation diffusion theory. The theory focuses on the 'process by which an innovation is communicated through certain channels over time among the members of a social system' (Rogers, 1983). This suggests that the four main elements in the diffusion process are the innovation, communication channels, time, and the social system. Research in this area has typically examined issues such as the time pattern of adoption, the categorization of adopters, and the individual adoption process. This literature usually characterizes the diffusion process by using a bell-shaped curve to describe the distribution of the time it takes to adopt a system among the members of a population. Rogers (1983) has classified the adopters of innovations into five categories based on the time relative to adoption: innovators, early adopters, early majority, late majority and laggard (as shown in Figure 1.2) (Rogers, 1983). Factors such as risk aversion, wealth, learning, imitation, and social pressure are often cited as the characteristics of adopters that influence the time of their adoption of an innovation. The growth pattern of the number of adopters

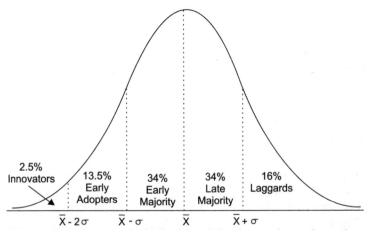

Adopter categorization on the basis of the relative time to adopt an innovation.

Figure 1.2   The Adopter Distribution
Note: $\bar{X}$ and σ refer to the mean and standard deviation of the adopter distribution respectively.

corresponding to the bell-shaped distribution is an S-curve. Such curves have been used successfully to describe the diffusion of many innovations.

Combining both the $O(N\log N)$ form of network value and the diffusion of innovation theories, researchers attempted to fit the growth of telecom networks and the Internet using S-curve models.

This network externality effect is especially strong for various telephone and computer networks as noted in Wang and Kettinger (1995), Gurbaxani (1990), and Sridhar (April, 2007) respectively.

There are typically two models which are used to model the growth of telecommunication and Internet services and are explained below:

LOGISTIC MODEL OF GROWTH

The approach to modelling the growth of such processes that exhibit the S-curve pattern assumes that the demand for access to communications is limited by a saturation level. It is also assumed that the rate of growth in the number of subscribers is positively influenced by:

1. the number of existing subscribers and
2. by the difference between the saturation level and the number of existing subscribers

The mathematical function that describes the above, popularly known as the 'logistic' curve is given below:

$$\frac{dy}{dt} = qy\left(y^* - y\right) \tag{1}$$

which can also be written as

$$y_T = \frac{1}{K + Ab^T} \, A > 0, \ 0 < b < 1 \tag{2}$$

where
y_t is the subscriber base at time t;
K, A and b are various constants.

Equation (2) represents an increasing S-curve which tends to the limit $1/K$ as $T \to \infty$. Equation (2) is non-linear and the various parameters can be estimated using a non-linear regression method.

The logistic model of growth has been used by Sridhar (April, 2007) in the prediction of mobile subscriber growth in India. (Also refer to Sridhar [31 October 2005] on the prediction of the saturation limits of the mobile subscriber base.) However, the forecasted predictions in this work were contradicted due to external environment conditions such as intense competition, reduced price of mobile handsets and other regulatory directives.

GOMPERTZ MODEL OF GROWTH

The Gompertz model has been widely used and has been found to be more appropriate for describing the growth of mobile services, especially in developing countries (Rouvinen, 2006; Singh, 2008; Sridhar, 2009). The Gompertz model is defined as follows:

$$MD_t = KA^{b^t} ; 0 \prec A \prec 1; 0 \prec b \prec 1 \qquad (3)$$

In (3), MD_t is the Mobile Density (number of subscribers per 100 population) at any time period t; K, A and b are constants. K defines the saturation level. The maximum growth rate is achieved when $MD_t = K/e$, that is, when mobile density reaches around 37 per cent of its saturation level (Gurbaxani, 1990). Sridhar (2009) estimated Equation (3) using the quarterly data of mobile subscribers from 1997–2008 and using non-linear least-square regression method. The estimates were made for a general Gompertz model with no upper bound on the saturation limit as well as with upper bounds of $K = 70$, 100 and 120. The Mean Absolute Percentage Error (calculated as $\frac{1}{|T|} \sum_{t \in T} \frac{(ActualMD_t - EstimatedMD_t)}{ActualMD_t}$) is minimal for $K = 120$. The actual mobile density and estimates using (3) with $K = 120$ are illustrated in Figure 1.3:

As can be seen clearly, there is a good overlap of the actual versus estimates until 2009, indicating a good fit. However, while the estimates indicate that the growth in mobile density would reach an inflexion point at a mobile density of 44 (at around the end of 2010), the departure of the actual mobile density curve from the estimates indicates continued growth. It is probably due to external factors that are discussed in the next section.

CRITICISMS OF S-CURVE MODELS

There are critics of the S-curve growth pattern, prominent ones being those who proved that Internet growth is still exponential and the saturation point as predicted by previous researchers has been surpassed (Rai et al., 1998). Specifically, the S-curve theories ignore such external factors as government policies, technological advancements, and service innovations. Sridhar and Sridhar (2006) and Sridhar (2 August 2005), point out that the saturation limit depends on a number of factors including growth in the disposable income of potential subscribers, price

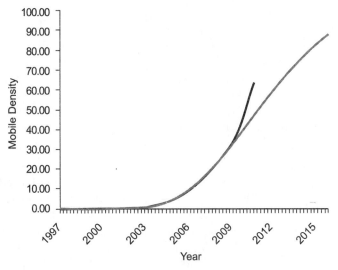

Figure 1.3 Modelling of Mobile Density in India using the Gompertz Model of Growth

of services, competition in the marketplace, price and availability of alternative communication channels, such as the Internet and regulatory policies regarding spectrum allocation and interconnection.

The simultaneous equation models that take into account all possible causal factors are difficult to estimate since available data is insufficient for a number of factors (Kathuria, Uppal, and Mamta, 2009). Further, it was shown in Sridhar (2009) that none of the above factors except network effect and competition level are the ones that significantly affect mobile subscriber growth in India.

Types of Telecom Services and Their Characteristics

Telecom which was the monopoly of the Indian government made a transition to a competitive market with the formulation of national telecommunications policies, and the formation of a regulator. A historical outlook and the forces that changed the government's intention, to allow entry of private operators in different segments are presented in the next chapter.

In general, all types of telecom services exhibit network externality characteristics and their adoption pattern conforms to the S-curve model described in the previous sections. However, the different services differ in the critical issues that govern their growth. Basic telecom services, commonly referred to as the wire-line service, exhibit high sunk costs and economies of density, more than any other type of service. Hence these were considered to be in the domain of a natural monopoly for a long time even in advanced countries such as the US, until technology advancements brought in competition. In India basic telecom service was provided by the Department of Telecommunications until competition was introduced. Even today, the government operators have more than 90 per cent of the market share of basic telecom subscribers of the country. Often considered as the only type of service to provide universal coverage, the importance of basic telecom services has decreased substantially in recent times, especially in developing countries, thanks to cheaper mobile services as an alternative solution to satisfy the telecom needs of the masses. These specific characteristics of basic telecom services with their unique cost structure, tariff regulation and their universal service obligations are discussed in Chapter 2 of the book.

Chapters 3 and 4 discuss the mobile services which propelled India to the second place in the telecom subscription base next only to China in a decade. However, the availability, allocation and management of the important resource for mobile services, namely radio frequency spectrum, has had a chequered and a very controversial path in the history of Indian telecom. Chapter 3 describes the technologies, market structure and licensing process of mobile services. The interconnection and quality of service regulation, mobile number portability and emerging virtual network operations are also explained. In Chapter 4, the different stages of spectrum allocation in India, including the recently concluded third-generation and broadband wireless spectrum are examined in detail. The nuances of spectrum management, especially the trade-off between competition and industry efficiency due to limited spectrum availability and spectrum fragmentation that is unique to the Indian context is also explained.

While Chapters 2–4 describe access networks, Chapter 5 discusses critical issues in two types of backbone networks, namely national long distance and international long-distance services. After competition was introduced in these segments, breaking the monopoly of the government monopoly operators, regulation of the interconnection of these backbone networks with the bottleneck access facilities assumed importance. Inter-carrier compensation requires the determination of optimal access charges. The framework for the access charges and the evolution of the

Interconnection User Charges regime in India are explained. In the case of international long distance service, India was under the International Accounting Rate regime until 2002 due to the monopoly operation of the government operator. The theoretical principles of international settlement are explained. The chapter concludes with the regulations and market structure relating to international cable landing and gateway operations explained through an important case in point.

The policymakers in India took a very liberal approach to Internet services by giving away licenses at a very low price compared to the millions charged for other telecom services. The Internet service is considered as a non-facility-based service in most of the countries as the service can be provided over the telecom infrastructure of the facility-based operators such as basic telecom operators. Even in India the pure play service providers remain non-facility-based service providers. However, the low barrier to entry facilitated by low licence fees did not help much in increasing the Internet user base in India. The Internet penetration in India is well below that of similar countries. One of the main reasons cited for low Internet penetration is the non-sharing of the last-mile access facilities by the incumbent basic telecom operators to Internet Service Providers, due to policy restrictions on 'Local Loop Unbundling'. The other reason is the restricted Internet Telephony regime adopted by the government due to non-level playing field between telecom companies and Internet service providers in terms of licence fees. Chapter 6 explains in detail the Internet service evolution in India starting with the erstwhile monopoly provisioning by the government operator. Restrictions on Internet service provisioning are explained and the chapter concludes with the most relevant topic of the day, namely the 'net neutrality' debate advocated by the Internet protagonists for removing any restrictions on content and carriage imposed by the basic or cellular mobile service providers. The chapter concludes with the development in the broadband infrastructure in the country and the associated policy initiatives.

Very Small Aperture Terminal (VSAT) satellite communication service that provides data connectivity across geographically distant locations has been one of the most regulated industry segments in India. The VSAT service comes under the ambit of 'Closed User Group' service. Satellite transponder restrictions, bandwidth restrictions, and restrictions on interconnection with public voice and data networks have restricted growth of this industry. Despite these limitations, VSAT services are used in certain niche user segments such as education, trading, and entertainment. The chapter on VSAT discusses the technologies, market structure and regulatory issues and illustrates the impact of these on the future growth of this segment.

While most of the above services are considered as telecom services under the Ministry of Communications and Information Technology, broadcast services such as cable television (TV) and direct-to-home (DTH) come under the purview of the Ministry of Information and Broadcasting. Cable TV, since its introduction in the early 1990s, has grown much like mobile services, thanks in part due to the unique registration-based local operator model. Cable TV services can be compared to basic telecom services as it involves the high sunk cost-based last mile wired local loop for provisioning of services. Chapter 8 discusses the unique market structure of cable TV services and the success or otherwise of the Conditional Access System (CAS) to bring in transparency in cable provisioning. Further, the alternatives to cable TV services, namely DTH and the recent advancements in other digital cable distribution services along with their regulatory implications are illustrated in this chapter.

Chapter 9 discusses first the hardware industry in the areas of network equipment, cables and transmission, hardware components, and handsets. The dramatic failure of domestic hardware manufacturing of the government-owned companies is explored, followed by the recent surge in the assembly and manufacturing of mobile user terminals and handsets. The Indian software industry has been actively engaged in the research and development of software outsourced services in the area of telecom equipment and handsets for foreign multinationals for over two decades. The growth of this outsourcing and the recent advancements in reverse outsourcing of software services by Indian telecos to multinational software firms are also briefly discussed.

The last chapter discusses partnerships in the telecom industry. The theoretical framework for mergers, acquisitions and other forms of partnerships is presented, and significant partnerships in the Indian and global telecom industry are mapped using the framework. The chapter also analyses the 'convergence' paradigm in telecommunications and its implications for consumers, telephone companies, service providers, and finally for policymakers and regulators. The concept of convergence in communications is presented and its impact on the evolution of the industry and the associated regulations are delineated.

The book provides a comprehensive review of all the sub-segments of Indian telecom, providing insights into critical technology, market and regulatory issues.

* * *

In this chapter we analysed the important characteristics of the telecom industry, namely network effects, and the various forecasting models that incorporate network effects in determining diffusion of telecom services. We concluded that most of the diffusion models do not take into consideration environment effects including regulations. An overview of various telecom services was given along with their specific characteristics.

2

Basic Telecom Services in India
From Monopoly to Oligopoly

Characteristics of Basic Telecom Services

The world over, voice communication is traditionally classified into three distinct types of services: local, national long distance and international long distance. Local telecommunication service, also referred to as Basic Telecom Service (BTS), described the provisioning of local access networks (normally referred to as the 'last-mile' connection) over relatively short distances. Each customer of the Basic Telecom Operator (BTO) normally has at least one last mile connection to the nearest exchange used to switch local, national long distance and international calls to the appropriate networks. Traditionally, this last mile connection is provided by the BTO using fixed wire-line consisting of a pair of copper cables and hence it is also referred to as a fixed-line service.

Basic telecom service was a 'natural monopoly' service in most countries. Until the late 1980s, the publicly owned national monopoly telecommunication companies were providing BTS in most of the European countries. In the US, until the Telecommunications Act of 1996 was implemented, BTS was being provided by monopoly private operators such as Regional Bell Operating Companies. India, like most nations, followed the government-owned monopoly PTT (Post, Telephone and Telegraph) operator model to start with.

The primary reasons for the natural monopoly form of telecommunication services have been huge fixed and sunk cost associated with the investment that makes duplication of the local loop uneconomical; network harmonization that is required for coordination and optimal utilization of various resources at points of interconnect; and the obligation to provide universal services. There are also certain strategic considerations such as the tendency of the governments to keep monopoly control over the telecom infrastructure so that resources can be mobilized against any threat to their sovereignty.

The natural monopoly factor is more pervasive and long-lasting, particularly in the rural telecom market due to the small size, low revenue potential, and slow growth prospects. Serving only limited rural localities may not be economically feasible for the service providers. Further, the rural population has limited capacity to pay due to the economic backwardness of rural areas. This results in a two-way problem: since economies of production are not fully realized, service cannot be provided at the lowest cost; at the same time, since the paying capacity of the target consumer group is low, higher prices cannot be charged for the service.

About 73 per cent of the global basic telecom markets were monopolies at the beginning of 1999 (InfoDev, 2000). Monopoly tends to restrict output, charge unnecessarily high tariffs, operate with bureaucracy and inefficiency and enjoys vast power. Liberalization, on the other hand, can mitigate these undesirable characteristics, and if widespread competition can be developed and maintained it can be expected to provide demonstrably superior results (Melody, 2000). The reasons for turning to privatization and liberalization in telecom have been consistent across countries. There was the realization that telecom was not a luxury but an essential service for economic development. It was also difficult to build such a vast infrastructure for telecom services without private sector capital (Chowdary, 1998).

History of Telecommunication Services in India

Telecommunication in India commenced operation in 1851 and has completed more than 150 years, to qualify as one of the oldest networks in the world. The Plain Old Telephone Service (POTS) also known as basic telephony, which was originally started by private companies, was taken over by the government in 1943. After independence, the Government of free India took charge of a network in 1948, which was quite small and connected only 82,000 subscribers in 321 switching exchanges with a total capacity of 100,000 lines. Since then, India continued the colonial legacy of the Indian Telegraph Act 1885 with telecom under the exclusive domain of the state until the mid-1990s.

However, very limited resources were allocated for the telecom sector as more priority was assigned to other sectors like roads, education, power, health, and so on. With no access to any sources of funds, for many years there was restricted growth in the telecom sector. Subsequent government plans allocated only limited resources for the telecom sector. Therefore, till 1985, resource deployment in the telecom sector was only around 2–3 per cent of the national plan outlay. At the beginning of the Seventh Five-Year Plan in 1985, the Indian government realized the need to improve the slow growth of the telecom sector. Hence, it constituted the Telecommunications Board and the Department of Telecommunications (DoT) within the Ministry of Communications to oversee operations, maintenance and development of telecom services.

A new emphasis was also placed on funding telecom projects. Keeping in view the importance of the sector, an increasing provision of outlays was made by the Indian government in the successive Plans. During the Ninth Plan and Tenth Plan, an outlay of Rs 47,280 crore and Rs 98,968 crore was allocated for the communication sector. The outlay for telecommunications in the different Plans is given in Figure 2.1.

Further, realizing that the development of world-class tele-communication infrastructure was the key to rapid economic growth and social change in the country, India embarked on the process of liberalization of the telecommunications sector in the early 1990s. The

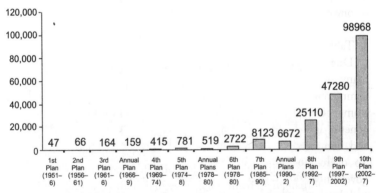

Figure 2.1   Plan Outlay for Telecommunications (in Rs crore)
Source: DoT, 2006.

world over the liberalization of the telecommunications sector was motivated by the following factors (InfoDev, 2003):

1. The need to attract private capital to expand and upgrade telecommunication networks and services,
2. Growth of alternative telecommunication services such as the Internet and mobile communication as a complement to fixed line services traditionally conceived to be natural monopoly services provided by the government,
3. Development of international telecommunication services provided by transnational and global telecommunication service providers.

In tune with the above, the Indian telecommunication sector underwent a major process of transformation through significant policy reforms, beginning with the announcement of the National Telecommunications Policy in 1994 (NTP 1994). The reforms were re-emphasized and carried forward under the New Telecom Policy in 1999 (NTP 1999). Driven by these various policy initiatives, the Indian telecom sector witnessed a complete transformation in the last decade.

In most of the countries, the process of privatization and liberalization started with three main activities:

1. Breaking the PTT structure and separating the telecommunication services from the postal services.
2. Formation of a credible regulatory system that can steer the industry towards a liberalized open market.
3. Corporatization of the monopoly operator and introduction of at least one competitor.

Table 2.1 presents the above three advancements in select countries.

Due to the public utility status that telecommunications held for many decades in developing countries, reforms geared towards the commercialization and privatization of services faced severe opposition from social groups with entrenched interests in the status quo. In general, labour, certain industrial groups, state managers, politicians, and some highly subsidized users would tend to oppose reforms. To overcome resistance from these politically powerful groups, the governments needed to gain a degree of insulation from civil society and develop a highly cohesive group of state officials in charge of reform implementation.

Table 2.1   Market Structure of Basic Telecom Services in Representative Countries

	US	UK	Australia	India
Initial model	Private Monopoly	Government Monopoly	Government Monopoly	Government Monopoly
Monopoly operator	AT&T	British Telecom	Telstra	Department of Telecommunication
Regulatory agency and when created	Dual: Federal Communications Commision (FCC) for Interstate; Public Utility Companies for intra-state regulation, 1932	Oftel (1984)	Austel,1989	Telecom Regulatory Authority of India (TRAI), 1997
Model subsequent to deregulation	Duopoly	Duopoly	Duopoly	Duopoly (1994)
Present model	Competitive market (1996)	Competitive market (1990)	Competitive Market (1997)	Competitive market (1999)

Source: Hudson, 1994; ITU, 2003; Jain, 2004.

Political Economy of Reforms in India

In June 1991 the Indian government launched a sweeping economic reforms programme that radically changed the structure of the country's economy. The reforms were initiated in response to the acute financial crisis the country faced in the aftermath of the foreign exchange crisis. In May 1994 the government announced that it would liberalize telecommunications as part of the overall economic restructuring.

Telecommunication services and the manufacture of telephone switches were deregulated for the first time in India in the 1980s. A DoT was created within the Ministry of Communications in 1985 to focus on enhancing the quality of telecommunication services. A year later two corporate entities were established:

1. Mahanagar Telephone Nigam Limited (MTNL) to provide local services in the metropolitan areas of Delhi and Mumbai, which

accounted for more than 20 per cent of the then installed fixed line subscriber base;

2. Videsh Sanchar Nigam Limited (VSNL) to exclusively provide international services.

The policy objectives in setting up MTNL were to:

1. introduce modern market-oriented professional management practices;
2. cater to the growth needs of industry and services in these two cities; and
3. establish a corporate model that would be replicated later in other parts of Indian telecommunications.

The new corporation was opposed, diluted and emasculated from within the Department of Telecommunications (Athreya, 1996). The DoT had been run for years by a permanent cadre of technical civil servants called the Indian Telecom Service (ITS). The majority of ITS officers, with honourable exceptions, saw MTNL as a dangerous erosion of their power. Political opposition from the DoT ensured that MTNL was not allowed to operate in any other Indian city. The Board of Directors of MTNL was not fully constituted. Mahanagar Telephone Nigam Limited was not treated as an autonomous financial entity. For all purposes of investment, recruitment, compensation, and pricing, Mumbai and Delhi were treated as just two more of the eighteen circles of Indian telecom. Some in the ITS and DoT hoped that in due course MTNL could be proved to be ineffective and absorbed back into the government departmental structure.

The Telecom Board, now referred to as the Telecom Commission was set up under the pursuance of Sam Pitroda in 1989 as a policymaking body within the DoT so that decisions could be prioritized and taken fast. The idea was to grant the Commission the financial and personnel powers then still exercised by the government. It was intended to be a 'government within the government' so that it could speed up decision-making and go for expansion. However, such hopes had evaporated by 1991. All proposals with financial implications still had to be referred to the Ministry of Finance. This covered any initiative on pricing, investments, imports, pay rises, and incentives. Similarly, all proposals with personnel policy repercussions elsewhere in the government had to be vetted by the departments concerned (Athreya, 1996).

Labour was a key factor that affected reforms in telecommunications in most of the less developed countries at that time. The then Prime Minister Narasimha Rao faced stiff resistance from more than 450,000 telecommunication employees of DoT who belonged to the National Federation of Telecom Employees and the Bharatiya Telecom Employees Federation for actively implementing the reform process (Petrazzini, 1996). Further, the Congress Party under Rao was also facing a threat from the Bharatiya Janata Party (BJP), which had gained considerable national support and threatened the reform process. The progress of reform headed by the then Finance Minister Manmohan Singh also suffered from lack of cohesion amongst the ruling Congress Party. Petrazzini (1996) cites that the permanent conflict among state officials created an institutional environment often described as a 'multiple-veto system'. He further goes on to argue that although technical and economic factors were important for analysing reforms in a country, it was primarily politics that affected the pace and shape of new telecom policies in India. Once reform was adopted as an official policy, the pace and viability of its implementation was largely affected by the degree of state autonomy and the cohesion of the governing elite, or in its absence, the degree of power concentrated in the head of the state. In India, neither of these factors was dominant throughout the years of reform (Petrazzini, 1996).

National Telecommunication Policies

The National Telecommunications Policy announced in May 1994 paved the way for the liberalization of telecom services and reaffirmed the need for the government to give priority to the development of telecom services in the country. The new telecommunications policy reflected the following views of the government (Sinha, 1996):

1. The development of telecommunications was vital to the success of wider economic reforms
2. That such development could not take place under the public monopoly model that was present in the country since independence in 1947.

The following section gives the objectives of the policy.

National Telecommunications Policy 1994

Following the objectives of the National Economic Policy and the Eighth Five-Year Plan, the NTP 1994 was formulated and the essential targets are listed below (Gupta, 2000):

1. Telephone should be available on demand by 1997.
2. All villages should be covered by 1997.
3. In the urban areas a public call office (PCO) should be provided for every 500 persons by 1997.
4. All value-added services available internationally should be introduced in India to raise the telecom services in India to international standards well within the Eighth Plan period, preferably by 1996.
5. The quality of telecom services should meet world standards.
6. It should be ensured that the defence and security interests of the country are protected.
7. Any company registered in India would be allowed to provide basic services. In mobile cellular and radio paging, the number of operators would be decided based on allocation of frequency.
8. As can be seen, the objectives set out in the policy were very broad to have any meaningful deployment guidelines. The tangible outcomes of NTP 1994 include the following:
    1. *Introduction of private sector participation in basic telecom services, thus ending the monopoly of government services.*
    2. *Introduction of second-generation mobile services with private sector participation.*
    3. *Creation of the TRAI in 1997.*
    4. *Liberalization of the VSAT policy to allow CUG services.*

New Telecommunications Policy 1999

The government recognized that the result of privatization was not entirely satisfactory after the implementation of NTP 1994. Except for the penetration of cellular mobile services in the cities, not much progress was made in the provisioning of basic telecom services by the private operators. As a result, some of the targets, including universal service obligation and telephone on demand remained unfulfilled.

In addition to some of the objectives of NTP 1994 not being fulfilled, and recognizing the convergence of telecommunications, the Internet and broadcasting, the Indian government drafted the New Telecom Policy (NTP) in 1999. The essential features of NTP 1999 are listed below:

1. Telephone density to be raised (from the present 2 per cent) to 7 per cent by 2005 and to 15 per cent by 2010.
2. Telephone-on-demand to be achieved by 2002.
3. Rural coverage for all villages by 2002 and rural teledensity to be raised (from the present 0.4 per cent) to 4 per cent by 2010.
4. National domestic long-distance services: Use of the existing backbone network of public and private power transmission companies, gas distribution firms, and Indian railways to be allowed for national long-distance voice and data transmission.
5. International voice telephony: The provisioning by monopoly government operators to be reviewed for opening up by the year 2004.
6. Corporatization of the DoT (a government department) which provided basic and national long-distance services by 2001 and separation of the departments into policymaking and service-provisioning arms.
7. Introduction of a universal service levy to provide corpus to the universal service obligation fund to compensate telecommunication companies for providing services in rural areas of the country.
8. Review of the Indian Telegraph Act 1885 and Indian Wireless Act 1993.

Regulatory Organization

Following the above liberalization and the entry of private operators in the provisioning of telecom services, the TRAI was created as a corporate body in 1997 through an Act of the Parliament (the Telecom Regulatory Authority of India Act 1997) to regulate the telecommunication services and to protect the interests of service providers and consumers. It is important to state that the reform process and the institutionalization of a regulator for telecom was also due to the impetus provided by the Uruguay round of the World Trade Organization (WTO) and the broad structural framework that the WTO reference paper on telecommunications required, to which India was a signatory.

ORGANIZATION AND STRUCTURE OF THE TELECOM REGULATORY AUTHORITY OF INDIA

The TRAI consists of a chairperson and not more than two fulltime members and not more than two part-time members to be appointed by the Central Government of India. While the chairperson and members constitute the TRAI, a secretariat assists the TRAI in the discharge of its functions. The secretariat is headed by a Secretary and is organized in the forms of divisions. The work of the TRAI is executed through these divisions, which number ten in all and are listed below:

1. Fixed network
2. Mobile network
3. Converged network
4. Economic
5. Financial analysis
6. Administration and personnel
7. Quality of service
8. Legal
9. Broadcasting and cable services
10. Regulatory enforcement

It is to be noted that the government, through a notification dated 9 January 2004, brought the broadcasting and cable television industry also under the umbrella of telecommunication services. The TRAI is heavily dependent on deputationists who constitute close to 50 per cent of the workforce. Deputationists are government employees, mostly working for the government-owned BSNL or MTNL, which seriously questions the independence of the TRAI from the government arms. It is often said that the TRAI is severely resource-constrained in its capacity to take up issues of importance and study them (Sridhar, 24 September 2003). Contrary to this, the Federal Communications Commission (FCC) in the US has six operating bureaus and ten staff offices, which deal with various functions.

Functions of the TRAI are captured in Figure 2.2 (TRAI, 2007).

The Telecom Regulatory Authority of India Act, 1997 was amended by the Telecom Regulatory Authority of India (Amendment) Act, 2000. The amendments were brought about to remove certain difficulties that had arisen in the implementation of the Act. The desired objectives of bringing about functional clarity, strengthening the regulatory framework and the disputes settlement mechanism were attained by bringing about a clear distinction between the recommendatory and regulatory functions of the TRAI by making it mandatory for the government to seek recommendations of the TRAI in respect of specified matters, by the setting up of a separate dispute settlement mechanism, and so on.

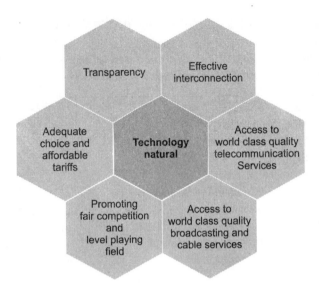

Figure 2.2   Functions of the Telecom Regulatory Authority of India

By the Amendment Act, an Appellate Tribunal known as the 'Telecom Disputes Settlement and Appellate Tribunal' was set up under Section 14 of the Telecom Regulatory Authority of India Act, 1997 by TRAI (Amendment) Act, 2000 (hereafter called the 'Act') to adjudicate on disputes and dispose of appeals with a view to protect the interests of the service providers and consumers of the telecom sector and to promote and ensure orderly growth of the telecom sector.

The functions of the Appellate Tribunal are to adjudicate on any dispute between a licensor and licensee, between two or more service providers, between a service provider and a group of consumers, and to hear and dispose of appeals against any decision or order of the TRAI. The Appellate Tribunal consists of a Chairperson and two Members. The Appellate Tribunal came into existence on 29 May 2000 and started hearing cases from January 2001.

FINANCIAL AUTONOMY OF THE TELECOM REGULATORY AUTHORITY OF INDIA

Section 11, clause (c) of the TRAI Act 2000 empowers the TRAI to 'levy fees and other charges at such rates and in respect of such services as may be determined by regulations' (TRAI, 2007). However, the options available to the TRAI are limited in terms of levying fees. Licensing not being the preserve of the TRAI, the major chunk of revenue garnered through licensing accrues to the licensor which is the DoT. Though the principal Act of 1997 did envisage the creation of a TRAI general fund under Section 22 of the Act, it has not been implemented. The Authority is fully funded by the grants from the Consolidated Fund of India that is routed through the DoT. In most of the countries licence fees cover the administrative cost of regulation (Sridhar, 22 September 2005). In case of FCC, even though the fees and charges it collects are deposited with the government, they are subsequently ploughed back into the Commission as budgetary support for its expenses.

DECISION-MAKING PROCESS

The decision-making process for critical telecom policy issues is as given in Figure 2.3.

It is often said that the TRAI's consultation process is very open and exhaustive. The consultation papers, stakeholders' response and the recommendations are distributed widely through email and the Authority's website. The various stakeholders including the service providers, equipment makers, handset makers, other service providers, and software firms are invited for giving their opinions during the open house, conducted mainly in the capital. However, once the recommendations reach the DoT, the opaqueness sets in. The discussions and documents at

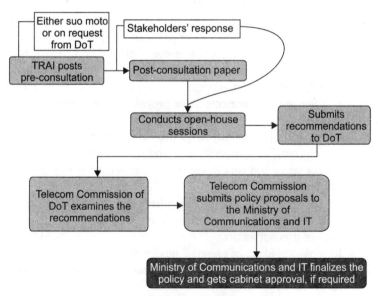

Figure 2.3   Decision-making Process for Telecom Policies

various stages of decision-making within the DoT are rarely made public. The policymaking decisions rest with the DoT and the Ministry.

The Telecom Commission was set up by the Government of India vide Notification dated 11 April 1989 with administrative and financial powers to deal with various aspects of telecommunications. The Commission consists of a chairman, four fulltime members, who are ex-officio Secretaries to the Government of India in the DoT and four part-time members who are Secretaries to the Government of India of the concerned departments. The functions of the Telecom Commission are indicated as below (Telecom Commission, 2010):

The Telecom Commission and the Department of Telecommunications are responsible for policy formulation, licensing, wireless spectrum management, administrative monitoring of PSUs, research and development and standardization/validation of equipment etc.

As indicated above, the government continues to be the administrator for policy and licensing. The functions of the TRAI are twofold: one recommendatory and another mandatory. As per the TRAI Act 1997, the functions and responsibilities of the TRAI are as follows:

1. Discharge its functions either *suo moto* or on request from the government in the following areas:
   a. Need and timing for introduction of new telecom service provider
   b. Terms and conditions of licence to the service provider
   c. Revocation of licence due to non-compliance of licence conditions
   d. Measures to promote competition and efficiency to promote growth of the telecom services
   e. Technology improvements in the provisioning of services
   f. Efficient management of available spectrum.
2. Discharge the following important mandatory functions:
   a. Ensure compliance of terms and conditions of licence
   b. Fix the terms and conditions of interconnectivity between service providers
   c. Lay down quality of service standards and undertake periodical audit of services provided by the service providers in the interest of users of telecom services
   d. Ensure compliance of universal service obligation
   e. Levy fees and other charges as determined by regulation.

LACK OF POWERS OF THE TELECOM REGULATORY AUTHORITY OF INDIA

Even though the TRAI has been constituted as an autonomous and independent regulator, the lack of policymaking and licensing powers confines its role to only that of a recommender most of the time. There were often turf-wars between the DoT and the TRAI. Instances in which there were confusion between the roles of the DoT and the TRAI are given below.

The root cause for the conflict lay in the original TRAI Act of 1997 which constituted the recommender and mandatory roles of the TRAI. The TRAI in its initial stages had interpreted that the provisions of the Act made it mandatory for the government to accept its recommendations even in areas where it had been given a recommender's role. Soon after its formation, the TRAI quashed the DoT's order of 29 January 1997 increasing the tariff on calls and denying two-way connectivity from fixed-to-mobile phones. The move by the DoT was thought to be detrimental to the growth of the just opened up cellular services sector. However, the DoT challenged the TRAI's order in the Delhi High Court which prompted the TRAI to ask for clarification of its jurisdiction and extent of powers on disputes between private operators and the DoT. Another case in point is the Calling Party Pays (CPP) issue, which is discussed in detail in the next chapter. The TRAI initiated the discussion of migration from the 'receiving party pays' scheme to CPP for cellular mobile calls, way back in its order of September 1999. But it was challenged in the Delhi High Court by the government-owned MTNL on the grounds that it did not have legal authority to implement CPP under the conditions of government-issued licensing (Sridhar, 24 September 2003). In January 2000, the high court struck down the recommendations of the TRAI on the grounds that the TRAI did not have the power to set licence conditions or revenue-sharing arrangements (Mukherji, 2004). In response to the Telecom Dispute Settlement Appellate Tribunal's (TDSAT's) directives on wireless local loop with mobility WLL(M) issues, the TRAI recommended to the DoT, the licensor, to curb the roaming facilities offered by the BTOs. But the Telecom Commission, the policymaking body of the government, reacted to it by saying that licensing conditions cannot be amended until both the licensor and licensee agree.

There was a controversy earlier between the TRAI and the DoT over Access Deficit Charges (ADC). When the TRAI was planning to revise

ADC, the Communications Ministry stepped in and argued that ADC was a policy issue and that the TRAI did not have the power to revise it. A similar incident took place over the TRAI's recent recommendation on spectrum allocation for mobile services. The TRAI's recommendation to form a Group of Ministers (GoMs) for allocating spectrum invited criticism from the Information and Communications Minister saying that the regulator had overstepped its mandate and provided recommendations outside its terms of reference. Though the TRAI's proposal for a GoMs to decide on spectrum allocation is questionable, it might have arisen due to the fact that it does not have direct jurisdiction over the functions of the DoT's Wireless Planning and Coordination (WPC) wing that looks after spectrum allocation issues (Sridhar, 22 September 2005).

The TRAI has been incapacitated as its role has been strictly limited to that of a recommender. Compare the TRAI's role with that of the FCC of the US. The FCC is an independent government agency directly responsible to the Congress. It is a regulator as well as a policymaker and licensor and hence is directly responsible for the enforcement of its rules, orders and authorizations. The Spectrum and Competition Policy Division which is responsible for formulating and implementing spectrum policies for wireless communications comes under the Wireless Telecommunications Bureau of the FCC. The Malaysian Communications and Multimedia Commission regulates as well as issues licenses. The National Telecommunications Regulatory, the Egyptian regulator, is empowered with spectrum allocation.

The need to empower the regulator for effective policymaking has been advocated by many (Sridhar, 24 September 2003; Sridhar, 22 September 2005). There are some instances in which the TRAI proactively demanded that the DoT and the Ministry should follow due diligence while making policy decisions. In a rare gesture to promote consumer interests, the regulator asked the DoT not to issue directives disallowing differential tariff for intra-network calls, way back in 2005. The DoT had earlier planned to issue a directive asking all the telecom operators to withdraw tariff schemes discriminating between calls terminating within the same network and calls terminating in other networks, which if implemented would have led to a hike in mobile tariff for millions of customers.

Though Section 11 of TRAI Amendment Ordinance (2000) has the following clause, it still leaves the final decision-making in the hands of the government:

The Central Government having considered that recommendation of the Authority comes to a prima facie conclusion that such recommendation cannot be accepted or needs modifications, it shall, refer the recommendations back to the Authority for its reconsideration, and the Authority may within fifteen days from the date of receipt of such reference, forward to the Central Government its recommendation after considering the reference made by the Government. After receipt of further recommendation, if any, the Central Government shall take a final decision.

Separating the roles of the licensor, policymaker and regulator complicates issues, especially with dispute settlement and modification of licensing conditions as the industry matures and evolves. If the government makes policies and licensing decisions, they mostly tend to favour the incumbent government-owned operators. For example, the DoT quashed the TRAI's recommendations on 'local loop unbundling' as it was against the interests of government-owned BSNL and MTNL.

The TRAI's role should also be strengthened from a purely recommending body to licensing and policymaking. Unless the regulator is empowered, the telecom landscape in the country will continue to be murky with distorted and vague policies.

Competition in the Basic Telecom Market

While most of the European countries and the US attained a fixed line teledensity of more than 50 (per 100 population) by 1995, India had a teledensity of 1.30. There were more than two million waiting to get a telephone connection (see Table 2.2 for the status of basic telecom services in India, 1992–7). Moreover, the government-operated telecom services in India had one of the highest fault densities of more than 200 faults per 100 lines. The monopoly government operator could not cope with the demand and improve quality of service due to limited funds and other resources. Hence with the declaration of NTP in May 1994, BTS were opened up as a duopoly market for private participation.

The decision to liberalize local services rather than the more attractive long-distance services as was the case in most of the countries including the US, was one of the most controversial decisions of the reform process (Sinha, 1996). Both within the government and in outside agencies such as the World Bank, there was the conviction that opening up of the basic services would not attract the private investment necessary to jumpstart the reforms. Many of the Regional Bell Operating Companies in the US such as Ameritech, South Western Bell, and Pacific Bell stayed away from participating in the bidding process since the financial viability of the basic services remained far from apparent. By liberalizing basic rather

Table 2.2   Status of Basic Telecom Services in India

	1992	1993	1994	1995	1996	1997
Main telephone lines in operation (in 000s)	6,797	8,026	9,795	11,978	14,543	17,802
Waiting list for main lines (in 000s)	2,846	2,497	2,152	2,277	2,894	2,706
Faults per 100 main lines per year (in per cent)	218.4	219.6	214.8	195.6	206.4	208.8

Source: ITU, 2003.

than long-distance service, the government disarmed potential criticism that it was not selling profitable business to private capital.

In NTP 1994, the government outlined an ambitious set of targets to be achieved through reforms: telephones on demand by 1997; all villages to be covered by 1997; and at least one PCO for every 500 persons in urban areas. These targets could only be met through massive investment in the expansion of the basic network. It was envisioned that the much needed investment would come from private operators.

Twenty-one service areas called 'circles' were formed for giving licenses. The twenty-one circles were further categorized in to A, B, and C based on their revenue generation potential, with Category A being the most prospective (see Table 2.3).

Table 2.3   Various Basic Telecom Licence Service Areas

Category A Circles	Category B Circles	Category C Circles
Andhra Pradesh (AP)	Haryana, Kerala	Andaman and Nicobar, Assam
Delhi	Madhya Pradesh	Bihar (including Jharkhand)
Gujarat	(including	Himachal Pradesh
Karnataka	Chhattisgarh)	Jammu and Kashmir
Maharashtra	Punjab, Rajasthan	North East
(including Mumbai	Uttar Pradesh (UP)	Orissa
and Goa)	West (including	
Tamil Nadu	Uttaranchal)	
	Uttar Pradesh (UP)	
	East, West Bengal	
	including Kolkata)	

Source: Gupta, 2000.

The creation of telecom circles for the award of licenses rather than a national licence, has strong social and political roots, and is closely related to the goals of universal service (Petrazzini, 1996). From the early stages of reform the then Prime Minister Rao and his administration were worried about the political consequences, and tried to do everything to avoid accusations of supporting a programme that benefited the rich and offered little to the poor. Since the government was seeking nationwide support for its reforms' programme, it feared that national licenses could have created an uneven development of the network. A national private operator could easily engage in cream-skimming, concentrating on states with good market potential (such as Maharashtra, Tamil Nadu, Gujarat) and ignoring the poor ones (such as Uttar Pradesh, Bihar, and the North Eastern states). Several of the latter (such as Uttar Pradesh and Bihar) have larger electorates, and in a country like India where the use of votes to punish unpopular political decisions is becoming widespread, a national licence for regional cream-skimming could have provoked a national electoral backlash. Hence the rationale for dividing the country into high, intermediate and low market potential. With the creation of circles based on the above categories, which closely resembled political divisions in the country, the government thought it had achieved a well-balanced telecommunications development programme for the entire country (Sinha, 1996).

The basic services' guidelines required that all exchanges and transmission systems should be digital; and all support systems be deployed for network and traffic management, billing, fault repair and customer services. There were some additional conditions imposed on bidders such as a licence period of fifteen years and minimum of 10 per cent of phones to be installed in villages for rural development. Stringent earnest money and high net worth requirements for bidding companies were enforced to avoid any truant bids and non-serious contenders. The Finance Ministry wanted the entire evaluation to be based on the financial bids submitted by the various bidders to maximize revenue for the government. However, the Communications Ministry wanted to give some priority to service requirements such as the speed with which the bidders would build their networks and the expansion of the network into rural areas (Sinha, 1996). Apart from the bid amount that carried 72 per cent weightage, the network rollout plan during the first three years carried 10 per cent weight; the percentage of those lines in rural areas carried a weight of 15 per cent; and the use of indigenous equipment carried 3 per cent weightage.

Though there was a big waiting line of customers who wanted to have telephone connections, the cost of deploying fixed line infrastructure was a major factor deterrent for most contenders. The cost of a wired access line ranged from Rs 25,000–30,000. Added to that was the socially desirable pricing adopted in most of the developing countries, in which the government set the tariffs only as much as the customers were able to pay (ICRA, 2002). Such a strategy dictated that the price of basic telephone service would be low, usually not related to the cost of provisioning the service. Hence it was less lucrative for new entrants to compete with erstwhile monopoly service providers.

With tariffs fixed at about Rs 200–50 per month by the government, the average period for break-even in the fixed line business was predicted at about seven to eight years (Jain, 2001). Further, line capacity was dedicated for each subscriber. If the capacity was not used, it would result in a dead investment. Hence the invitation for a BTS licence received only a lukewarm response. After a long delay and series of disputes, licenses were given to private operators to compete with government monopoly operators in six service areas of the country, leaving fifteen other circles to the monopoly regime due to the high bid amount, uncertainty of the market potential and certain legal issues. Table 2.4 gives details of the first round of licensing in the six telecom circles.

Introduction of Competition in the Basic Telecom Market

By 1999, of the six private basic service operators only two had commenced service in the country. Since most of the private basic service operators

Table 2.4   First Round of Basic Telecom Licensing and Corresponding Licence Fees across Licence Service Areas

Operator	Licence Service Area	First Year Licence Fee (in Rs Crore)	Total Licence Fee (payable over 15 years) (in Rs Crore)
Bharti	MP (B)	19.20	654.50
Shyam Telelink	Rajasthan (B)	10.00	1,110.00
Reliance Telecom	Gujarat (A)	97.04	3,396.00
Tata Tele-Services	AP (A)	120.50	4,200.00
Essar Commvision	Punjab (B)	131.24	4,593.40
Hughes Ispat	Maharashtra (A)	397.40	13,909.00

Source: Gupta, 2000.

were not able to fulfil their commitment towards the licence fees, on recommendations from the Attorney General of India, the Government of India shifted all the licence holders to a revenue-sharing scheme. Under this scheme, the BTS licence holders would pay annually 6–10 per cent of their gross revenue as licence fee to the government. However, in return for the bailout package, the operators had to give up their duopoly status.

The New Telecom Policy enacted in 1999 (NTP 1999) proposed a higher degree of competition in the basic services segment for the circles yet to be licensed. It envisaged multiple operators during the transition period of five years. With recommendations from the TRAI, policy guidelines for allocating additional basic service licenses were issued by the Government of India in January 2001. The guidelines allowed the BTOs to use Wireless Local Loop (WLL) technology as well for faster deployment. Recognizing that the universal service obligation might not be fulfilled under normal commercial considerations, NTP 1999 stipulated the creation of a Universal Service Obligation (USO) Fund through the imposition of a universal access levy as a percentage of the revenue earned by the operators under various licenses. The universal access levy, required for supporting the provision of village public telephone and rural exchange lines would cover both capital expenditure and recurring expenses to run the services. Applicants who met the minimum requirements on financial net worth and paid-up equity capital as laid down in the licensing guidelines had to pay a stipulated entry fee and provide bank guarantees to get the licence.

In the second round of licensing during 2001, the government decided to further open up basic services without any restrictions on the number of entrants. The licenses were issued to applicants who satisfied the minimum paid-up equity capital and minimum network requirement specified for each circle (refer to DoT, 2001 for details on prescribed eligibility criteria for BTS licenses). The applicants were required to pay a one-time prescribed entry fee for each circle. Performance bank guarantees were to be submitted by the licensees which would be released after due completion of prescribed rollout obligations. In addition to the entry fee, the licensees had to pay the prescribed annual licence fee as a percentage of Adjusted Gross Revenue (AGR) (see Table 2.5 for details on the entry fee).

However, despite the entry of private operators, the basic wire-line service has registered only moderate growth over the years and has started showing decline due to the growth of mobile services.

Table 2.5 Licence Fee for Second Round of Basic Telecom Services Licensing

Telecom Circles	Entry fee (in Rs Crore)	Percentage of Revenue as Licence Fee
Category A Circles		
Andhra Pradesh	35.00	12
Delhi	50.00	12
Gujarat	40.00	12
Karnataka	35.00	12
Maharashtra (including Mumbai and Goa)	115.00	12
Tamil Nadu (including Chennai)	50.00	12
Category B Circles		
Haryana	10.00	10
Kerala	20.00	10
Madhya Pradesh (including Chhattisgarh)	20.00	10
Punjab	20.00	10
Rajasthan	20.00	10
Uttar Pradesh (West) (including Uttaranchal)	15.00	10
Uttar Pradesh (East)	15.00	10
West Bengal (including Kolkata)	25.00	10
Category C Circles		
Andaman and Nicobar	1.00	8
Assam	5.00	8
Bihar (including Jharkhand)	10.00	8
Himachal Pradesh	2.00	8
Jammu and Kashmir	2.00	8
North East	2.00	8
Orissa	5.00	8
Total	**497.00**	

Figure 2.4 shows the growth of basic wire-line services.

Wireless Basic Services

Traditionally, basic services were being provided using copper-based wired local loop connecting the subscriber premise to the end-office of the Public Switched Telephone Network (PSTN). In India, the cost of the wired local loop was very high and was touted as one of the main

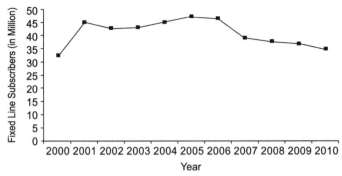

Figure 2.4   Trends in Basic Wire-line Services in India

hurdles in implementing a cost-effective local loop access, especially in rural and remote areas (for details on the cost of BTS, refer to Box 2.1).

However, the basic services guidelines allowed the use of WLL as follows:

Basic service operator shall be allowed to provide mobility to its subscribers with wireless access systems limited within the local area that is, Short Distance Charging Area (SDCA) in which the subscriber is registered (also called as WLL(M)). While deploying such systems, the operator has to follow numbering plan of that SDCA and it should not be possible to authenticate and work with the subscriber terminal equipment in SDCAs other than in which it is registered. The system shall also be engineered so as to ensure that hand over of subscriber does not take place from one SDCA to another SDCA while communicating. (DoT, 2001)

It is also clarified in the guidelines that there will not be any extra one-time charge for the acquisition of 5+5 MHz radio spectrum for WLL(M) services. The spectrum holder has to pay appropriate annual spectrum charges. In line with the above, the Telecommunications and Computer Networks group at the Indian Institute of Technology (IIT) Chennai under the leadership of Professor Ashok Jhunjhunwala took the initiative to develop an indigenous cost-effective fixed wireless local loop solution called corDECT (Jhunjhunwala, 2000). Apart from reducing the per line cost to Rs 12,000–15,000, corDECT provided both voice and Internet connectivity. Highly appreciated by the UNDP as a fast and cheaper mode of accessing the Internet in developing countries, corDECT was adopted by BTOs such as BSNL, Shyam Telecom and Tata Teleservices, especially for rural areas in their respective circles.

At the same time, following the above guidelines, some of the BTOs (namely Tata Teleservices and Reliance Communications) approached

Box 2.1 Costs in Providing Basic Telecom Services

A typical local loop plant of a BTO is shown in Exhibit 1. The subscriber access loop (indicated as (3) in the Exhibit) consisted of a pair of twisted pair cables bundled together in a local access loop (1) which was connected to one of the remote exchanges (2) in the service area. The remote exchanges in a city or a service area were interconnected through a high-bandwidth multiplexed self-healing ring architecture (4) to the main exchange using synchronous digital hierarchy (SDH) or Metro Ethernet (MEN) technologies.

The largest single cost of providing telephone service is the cost of the local loop that connects a customer to the first point of switching in the telephone network. It is estimated that this cost is close to two-thirds of the cost of the network. The cost of a local loop is fixed and it does not vary with the number or duration of calls that are placed or received.

There are various cost elements involved in the provisioning of fixed line service by the BTOs:

Local Access and Subscriber Access Cost: These costs are related to deploying twisted pair cables from the remote exchange to the subscriber premise indicated as (1) and (3) in the Exhibit. These include costs of copper cables; cost of digging and restoration of roads, highways and paths; and fees paid to various agencies for Right of Way (RoW) for laying down the local loop. Also included are the costs of subscriber premise wiring and the cost of instrument (if rented). These are sunk costs and hence once incurred cannot be recouped if the new entrant ceases service. The need to incur these substantial upfront sunk costs constitutes a significant barrier to entry. When an incumbent has already deployed sunk facilities to serve all customers, a new entrant might be unwilling to sink the costs of duplicative facilities, either because it might be unable to lure customers away from the incumbent and generate enough revenue to cover those substantial sunk costs quickly, or because the resulting competition between itself and the incumbent would drive prices so low that even if the new entrant achieved a significant market share, it would still be unable to recover its sunk costs.

Exchange Infrastructure Costs: These include the cost of renting/leasing buildings for locating the main and remote exchanges, indicated as (2) in the Exhibit.

Transport Facility Costs: As indicated in the Exhibit, the remote and main exchanges are typically interconnected using a transport ring network (4). This again requires digging, restoration, and RoW across the service area/city and hence the costs associated with these. Further, the cost of optic fibre links that are used in this transport network also needs to be included.

Transport Electronics Cost: This is associated with the cost of electronics ((5) in the Exhibit) needed for SDH or MEN housed in the remote/main exchange for transport of calls.

Switching Cost: This is associated with the cost of electronics needed in the exchange for switching voice/data/video ((6) and (7) in the Exhibit).

Exhibit 1—Basic Service Operator's Network Architecture

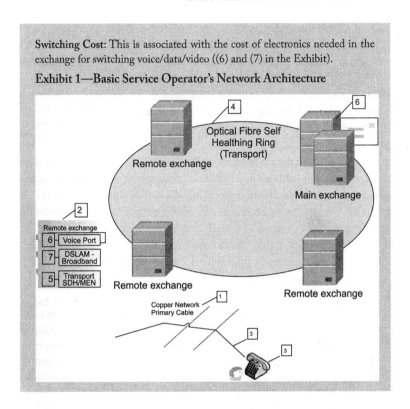

the government with a proposal that they could provide a local access loop at a much lower cost using CDMA technology. The CDMA technology, developed by QualComm in the US as a competing technology to GSM for providing cellular services found acceptance in emerging markets. The BTOs argued that quick deployment along with high spectral efficiency and lower per line cost of less than Rs 10,000, definitely made CDMA a better alternative compared to wired access loop for certain areas of the country. The technology being mainly developed for cellular services could also provide subscriber mobility. This move blurred the distinction between basic services and cellular mobile services. A new type of service called 'Wireless Local Loop with Limited Mobility' (WLL-LM) (that is, limited to be mobile within the SDCA as per the basic services licence guidelines) came into existence.

However, the WLL-LM services proved to be a strong competitor for cellular mobile services. Since the percentage of intra-SDCA calls on an average was around 80 per cent compared to the 20 per cent inter-SDCA calls, and the intra-SDCA calls were offered at the low price of basic services by the WLL-LM operators, most of the traffic on the networks of Cellular Mobile Service Providers (CMSPs) shifted to the network of WLL-LM operators, thus causing concern to the cellular operators. The ministerial Group on Telecom and IT (GoT-IT) committee to whom the controversy was handed over referred the matter to the TRAI and TDSAT. However, the committee noted that 'if technology allows limited mobility as an extended service to the advantage of the customers then it should be permitted'. Regarding spectrum allocation, the committee recommended the 'beauty contest' approach, wherein the spectrum would be allocated based on the rollout plans of the BTOs and that the operators need not pay any licence fees for the spectrum allocated for LM (Sridhar, 4 January 2001).

After many legal battles, in November 2003, the Indian government announced the guidelines for Unified Access Service Licence (UASL) which allowed the use of any technology (both wire-line and wireless) for the provisioning of network access service, with an option for BTOs to migrate to UASL by paying the prescribed fee. The entry fee for migration of a BTO licence to the UASL for a service area was set equal to the entry fee paid by the Fourth Cellular Operator for that service area (refer to Chapter 3 for details on entry fee), or the entry fee paid by the BTO itself, whichever was higher. While applying for migration to UASL, the BTOs would pay the difference between the said entry fee for UASL and the entry fee already paid by it. Most of the BTOs migrated their licence to UASL in different circles. This ended the controversial era of the WLL-LM service during 2001–3.

The government further put restrictions on the provisioning of fixed wireless service as the technology allowed the subscribers to roam with the equipment within the coverage area of the access base station. This functionality being not in consonance with the principles of basic services, the following amendments were made to the provisioning of fixed wireless basic services (DoT, 2005):

Terminal used for fixed wireless services should be strictly confined to the premises of the subscriber where the telephone connection is registered. It should also be noted that it is licensee's responsibility to ensure that the subscriber terminal is operated in accordance with the terms of the License for fixed lines including this clarification.

This led to two types of services that could be provided by the BTOs: (i) fixed wire-line service and (ii) fixed wireless service with the terminal confined to the subscriber premise where it is registered. However, by December 2009, the fixed wireless service provided by the various operators accounted for less than 10 per cent of the total number of fixed lines provided by the BTOs.

Pricing of Basic Services

There are two possible alternatives for pricing any telecommunication service: (i) socially desirable pricing and (ii) cost-based pricing. The principle of socially desirable pricing, which forms the basis for basic telecom pricing strategies in most of the developing countries, is to charge customers only as much as they are able to pay. Such a strategy dictates that the price of basic telephone service would be low, usually not related to the cost of provisioning of service. The rental and usage charges are normally cross-subsidized by the more expensive business, long-distance, and international long-distance services. In a cost-based pricing structure, the tariff is based on cost of the network elements in providing service. Fixed charges, called rentals, recover the capital costs (CAPEX) of the infrastructure required to build the network. Variable charges, known as usage charges, recover operation, maintenance, administrative and interconnect charges (OPEX), based on minutes of usage. Until competition was introduced in basic services, subscribers of the erstwhile government operator were paying very high long-distance and international call charges to cross-subsidize basic services. This is often referred to as 'service-service inequity'.

However, the move from socially desirable pricing to cost-based pricing, especially in an evolving telecom market such as that in India is difficult. Various tariffs for basic telecom services in India are shown in Table 2.6. A look at the charges shows little variations from 1999 until 2003, which is indicative of the tariff-fixing policy of the government and the regulator for basic telecom services. The usage charges are per minute unlike the flat charges charged in most of the developed markets. The rentals are low compared to other countries.

INTERCONNECT CHARGES

Economic and technical advantages of an incumbent are due to the network externality effect. For their networks to attract any subscribers and offer value, new entrants must interconnect with the incumbent's network so that their customers can communicate with the large subscriber base of the incumbent. Denial of access to the incumbent's network would harm the new entrants. Without interconnect to the

Table 2.6   Basic Telecom Charges in India

		On 31 March 1999 (in US$)	On 1 April 2003 (in US$)
Rural Rent Per Month	Exchange Capacity		
	<1,000	1.18	1.48
	1,000–30,000	2.35	1.48
	30,000–100,000	3.23	2.53
	100,000–300,000	4.23	4.21
	300,000+	4.47	5.90
Rural Call Charges	First 200 metered calls	0.01	5.90
	Next 50	0.02	0.02
	Next 50	0.02	0.02
	Next 200	0.02	0.02
	Next 250	0.02	0.03
	Next 1000	0.03	0.03
	Further Calls	0.03	0.03
Urban Rent Per Month	Exchange Capacity		
Non-commercial	<100	1.18	2.53
	100–999	1.76	2.53
	1,000–30,000	2.35	2.53
	30,000–100,000	3.23	4.21
	100,000–300,000	4.23	5.90
	300,000+	4.47	5.90
Commercial	30,000–100,000	Not Available	4.64
	100,000+	Not Available	6.53
Urban Call Charges	First 175/Month	0.02	0.02
	Next 125	0.02	0.02
	Next 125	0.02	0.03
	Next 75	0.03	0.03
	Next 425	0.03	0.03
	Further Calls	0.03	0.03

Source: Desai, 2006.

incumbent's network, the calling opportunities for the new entrants' subscribers to other fixed telephone subscribers would be less than 1 per cent of the total calling opportunities that arise with interconnect to the incumbent's network (ICRA, 2002). However, since there are so few subscribers on the new entrant's network, neither the incumbent nor its subscribers would reasonably attach any value to being interconnected to the new entrant's network. Since the incumbent currently served virtually all subscribers in its local serving area, it had little economic incentive to assist new entrants in their efforts to secure a greater share of that market. Even with interconnect, the erstwhile incumbent monopolist could discriminate in providing interconnect in various ways: by providing poor interconnections; slowly and/or ineffectively repairing and maintaining leased inter-network facilities; charging high prices for interconnect; and in general imposing other onerous conditions for terminating calls from the new entrant's customers to the incumbent's subscribers. It is natural for an incumbent monopolist to have strong incentives to behave this way, leading customers to reject alternative providers and so preserve local monopolies. High interconnect prices could then force the entrant to suffer financial losses. Inadequate and poor interconnect prevent the entrant from competing effectively for customers who care about service quality. In most of the countries including India, the regulator stepped in with appropriate polices to prevent such erstwhile monopoly incumbent's abuses (TRAI, 2003).

On 1 May 2003, the Telecommunication Interconnection Usage Charges (IUC) regulation notified by the TRAI came in to existence (TRAI, 2003). In this order the TRAI tried to marry the above two pricing strategies. The IUCs were specified in the following three categories:

1. Termination Charges: Termination charges for calls to basic (fixed, WLL (fixed), WLL (mobile) and cellular mobile networks were specified at a uniform rate of Rs 0.30 per minute.
2. Origination Charges: Forbearance and hence as dictated by the market.
3. Access Deficit Charge: Charges to be paid by BTOs, mobile operators, national long-distance and international long-distance operators to the BTOs for certain types of calls to cover the deficit between the cost-based monthly rental and average monthly rentals.

ACCESS DEFICIT CHARGES

The TRAI estimated the cost-based monthly rental for wire-line based basic telecom services to be Rs 424, which should have been borne by the subscribers if the TRAI had adopted cost-based pricing. However, in order to make the basic services affordable to everyone, the TRAI fixed minimal increase in rentals and usage charges. An ADC was computed that was to be paid to the BTOs who provided fixed line service (wire-line or wireless) to cover the deficit between the cost-based monthly rental and average monthly rentals. Thus ADC is the amount payable by the service provider at the caller's end to the service provider at the receiving end (only the BTOs) for accessing services rendered by the latter in terminating cellular mobile, domestic long-distance, and international long-distance calls. Table 2.7 details the ADC implemented by the TRAI in 2003. As can be seen in the table, ADC was levied on non-basic services such as mobile (either termination or origination), long-distance, and international long-distance calls. The WLL-LM services provided by certain fixed-line BTOs did not attract ADC. However, TDSAT ruled against this and said that the CDMA operations of the basic service providers would not be viewed as fixed service for the calculations of ADC (Mukherji, 2008).

However, economists argued that ADC had to go and the BTOs had to adapt to cost-based pricing in a competitive market. Realizing that such cross-subsidization was not long-lasting, ADC charges were continuously decreased by the TRAI over time. The minute-based ADC was abolished for domestic calls and the ADC was set at 1.5 per cent of the adjusted gross revenue in 2006 (TRAI, 2006). The TRAI further noted that ADC would be merged with the Universal Service Fund (discussed in the next section) and that it should be reduced to zero by 2008–9. The ADC on all calls except international incoming calls was reduced to zero by March 2008. The ADC on outgoing international

Table 2.7   Access Deficit Charges

Type of Call	Local	Intra-circle Calls		Inter-circle Calls			International Long-distance (in Rs)
		0–50 km	>50 km	0–50 km	50–200 km	>200 km	
Fixed to Fixed	0.00	0.00	0.30	0.30	0.50	0.80	4.25
Fixed to Cellular	0.30	0.30	0.30	0.30	0.50	0.80	4.25
Cellular to Fixed	0.30	0.30	0.30	0.30	0.50	0.80	4.25
Cellular to Cellular	0.00	0.00	0.00	0.30	0.50	0.80	4.25

Source: TRAI, 2003.

calls was reduced progressively from Rs 2.50/minute to Rs 0.80/minute. On incoming international calls, the ADC levy was reduced from Rs 3.25/minute to Rs 1.60/minute, to Rs 0.50/min and finally to zero in September 2008. Mukherji (2008) presents a detailed political economy perspective on the phasing out of ADC in favour of GSM operators, thus contributing to a surge in foreign investor confidence in Indian telecom. Box 2.2 describes some of the effects of ADC from the public policy point of view.

Box 2.2   The Substitution Effect on Access Deficit Charge

An ADC was payable to the basic service providers (mainly the government-owned BSNL) on a per minute basis by operators of cellular, unified access, national and international long-distance services. The TRAI introduced the ADC in 2003, to compensate the basic service providers for providing high-cost fixed-line connectivity at low rentals, especially in the rural areas of the country. The ADC for incoming international calls to a mobile phone was set at Rs 3.25 per minute compared to Re 0.30 for a local call. This huge difference enticed certain international long-distance operators to masquerade an international call as a local call, thus saving about Rs 2.95 per minute of call usage.

What are the implications of ADC from a public finance point of view? First is the substitution effect. The ADC is analogous to a commodity-specific tax, being charge on a specific telecom service (international long distance). Assuming that the demand for international phone calls is elastic and assuming at first, that companies are honest and pay the charge, the imposition of ADC will make such calls more expensive for consumers. In this case, incentives exist for the consumer to substitute international calls with cheaper alternatives such as electronic mail, chat, or, Internet telephony call services, available, for example, from Internet Service Providers (ISPs). In all such cases, where possibilities of substitution exist, public finance shows that the imposition of ADC can cause deadweight loss, since there is a loss to the consumer, who substitutes international calls with the relatively cheaper, lesser quality service such as Internet telephony. Since, in this case, the ADC-levied service is not being used, the revenue earned by the government operator is also reduced.

The second effect of the ADC is the evasion of these charges by the operators through masquerading. There could be several reasons for companies to be able to offer low prices on international calls, of which non-payment of ADC is just one, sly alternative. There is no doubt that the consumer making international calls will stand to save. But the government operator stands to lose some part of revenue from ADC, when masquerading exists.

Now it can be shown and extended from the standard public finance theory that if the ADC has the effect of creating a deadweight loss, its non-payment will add to this cascading effect. This is especially so if the revenue loss to the government from non-payment of the ADC were to be higher than the savings for the consumer due to a low price for international calls (assuming the low price is attributable to non-payment of the ADC).

Based on standard public finance, it can be shown that an amount equivalent to the ADC, if imposed as a lump-sum charge (rather than as per minute ADC on specific services such as international calls) on telecom companies, increases revenues. An example is the revenue share of 5 per cent of adjusted gross revenue levied on all telecom service providers (except ISPs) towards the USO Fund. Such a lump-sum charge will cause no consumer substitution that would result from a commodity or service-specific charge.

The thinking behind this is that with a lump-sum charge on a larger revenue base of the companies, higher revenue can be generated, than with a service-specific charge, while allowing consumers to use their preferred service (not substitute cheaper alternatives). This may not make sense in one time period, but it does make a lot of difference to revenues from year to year, when profit bases of companies are reduced due to the specific charge.

The final move by the TRAI to merge the ADC with the USO Fund removes any distortion caused by economic inefficiencies.

Universal Services

There are wide differences of opinion over the definition of universal services. There are various factors such as historic, legal, philosophic and cultural that could help explain such differences in approach. However, the general consensus on the definition of universal service includes:

Availability: Provision of telephone service, whenever and wherever required even in rural and remote uneconomic areas.

Affordability: To bring the standard price of basic service within everyone's reach or targeting special low-cost service at low-income groups.

In most countries, rural coverage and universal service provisioning was a licensing requirement for basic fixed-line service providers. Countries including India created a special government fund, normally referred to as the 'Universal Service Obligation Fund', to meet the specified universal service obligations.

What should be the goal of policymakers—universal service or universal access? Universal service that is, 90 per cent of homes having a telephone is neither a warranted goal nor desirable. Universal access is socially desirable as the benefits of science, technology, and telecommunications are extended to whosoever wants them, without much efforts, and are affordable, whenever needed (Chowdary, 1998).

The most important factor to govern the growth of telecom and especially universal service is affordability. In developed countries like the US, 90 per cent of the total households can afford a monthly expenditure of US$ 30 per month that is, 5 to 7 per cent of the total monthly income on communications (Jhunjhunwala, 2000). Thus US$ 360 per year per household return is sufficient to pay for cost of capital, operation, maintenance and obsolescence cost, and is capable of garnering a good amount of profit as well. The situation in developing countries such as India is quite different. Table 2.8 shows the distribution of households under different income brackets and the telecom expenditure that they can afford.

From Table 2.8 it is clear that only 1.6 per cent of the households can afford the annual expenditure of US$ 360 or more. In India the average household income in urban and rural areas is Rs 60,000 per year and Rs 40,000 per year respectively. The rental alone in urban and rural areas is Rs 2,400 (4 per cent of the annual income) and Rs 840 (2.1 per cent) respectively. In universally connected developed countries, the corresponding proportions are 0.24 per cent and 0.3 per cent respectively.

The USO regime under NTP 1994 required 10 per cent rural coverage as a licensing requirement for basic fixed-line service providers. However, most of the BTOs were not able to fulfil their rural rollout obligations. Cross-subsidization of low rental and usage charges also would not work in the long run. Hence NTP 1999 envisaged an entirely different approach to implementing USO, charging Universal Service Levy (USL) at a prescribed percentage of the revenue (5 per cent in 2007) earned by the operators holding different types of licenses. Basic service providers who fulfilled their USO for rural and remote areas would be reimbursed

Table 2.8   Income and Affordability of Telecom Services in India

Annual Household Income (US$)	Affordable Telecom Expenditure (in US$)	% of Households
>5000	>360	1.6
2500–5000	175–360	6.3
1000–2500	70–175	23.3
500–1000	35–70	31.8
<500	<35	37

Source: Jhunjhunwala, 2000.

the net cost of providing the universal service from the funds collected by way of USL. The TRAI recommended setting up an independent authority for the administration of the Universal Service Fund.

The government operator BSNL continued to take the responsibility of installing Village Public Telephones (VPTs). However, the rural teledensity in the country was very low at 1.94 per 100 population compared to 31.13 of the urban areas (TRAI, 2005). The TRAI made a series of recommendations including considering USO subsidy for mobile services (TRAI, 2005). The Indian Telegraph (Amendment) Rules 2006 were subsequently modified in order to enable support for mobile services and broadband connectivity in rural and remote areas of the country. The progress since then on rural teledensity has been very good. The accumulation of the USO fund since it was levied is shown in Table 2.9.

Local Loop Unbundling

One of the main goals of liberalization had been to encourage the building of new infrastructure as a complement to the incumbent's to meet the increasing demand. The financial burden and risk of deploying competitive networks outside key urban business districts were so high that the number of new entrants willing to expose themselves to the risk of constructing nationwide access networks remained limited. The total gross block of the government operators' telecom network was in excess of US$ 20 billion at the end of March 2001. The new entrants seemed to

Table 2.9 Allocation and Distribution of Univeral Service Obligation Fund as on March 2010

Financial Year	Funds Collected as USL (in Rs Crore)	Funds Disbursed (in Rs Crore)	Reimbursement of Licence Fee and Spectrum Charges (in Rs Crore)
2002–3	1,653.61	300	2,300
2003–4	2,143.22	200	2,300
2004–5	3,457.73	1,314.59	1,765.68
2005–6	3,215.13	1,766.85	582.96
2006–7	3,940.73	1,500	
2007–8	5,405.80	1,290	
2008–9	5,515.14	1,600	
2009–10		2,400	
Grand Total	25,331.36	10,371.44	6,948.64

Source: USO, 2011

be in no hurry to spend huge amounts to build facilities that duplicated the facilities of the government operators. Replicating even a portion of the existing local telephone network was expensive, and the business risk was high. Even if operators were keen to finance fresh investments, construction time would have run into years. The incumbents enjoyed advantages of economy of operations which made it very difficult initially for even the most efficient of new entrants to compete.

Further, the Indian Telegraph Act, 1885 allowed the incumbent government operators the RoW in the use of public space, and rights to enter private lands and buildings. However, new entrants had to make their own arrangements for RoW. They also faced major implementation delays because of difficulties in obtaining RoW from various agencies. The TRAI had insufficient power to ensure RoW for new entrants. Such power was vested with the Ministries (such as the Ministry of Surface Transport and the Ministry of Railways) of the Government of India, and with municipalities. The situation was complex, with varied ownership arrangements, various cost-sharing arrangements among users, and circumstances in which different terms of access and rates applied to incumbents and new entrants. This complicated the process of obtaining RoW and resulted in increased cost to the service providers (Sridhar and Malik, 2007).

For the above reasons, BTS continues to remain a virtual monopoly of the incumbent operators. There are about 48 million fixed lines in India, out of which the incumbents own more than 80 per cent.

Way back in April 2004, the TRAI had suggested non-discriminatory Local Loop Unbundling (LLU) so that the access networks of the BTOs could be shared with ISPs and other competing operators for improving broadband penetration in the country (TRAI, 2004) (refer to Box 2.3 on various types of LLU). However, the incumbent operators viewed LLU as infringement of their property rights and hence had prevented the government from implementing it.

Though success rate of LLU in other countries are still being debated, in a competitive Basic Telecom market, LLU as a regulatory intervention is seen as an instrument to discipline market power, reduce monopolistic bottlenecks and possibly provide way for innovative service offerings such as broadband connectivity. Most of the broadband connections are in metros and large cities. However, penetration of broadband is poor in smaller cities and outside urban areas where only government operators are providing fixed line service. As has been pointed out earlier, it is often uneconomical for private operators to lay down new lines in rural

Box 2.3   Local Loop Unbundling

Unbundling the local loop can take many forms. At its simplest, unbundling may involve shifting the point of interconnection for switched services down to the trunk side of the local exchange. More extensive forms of unbundling involve the supply of separate switching and transport elements at the different network levels, the provision of signalling services separately from carriage services, or access to the Intelligent Network functionality used in the local network, such as the 1-600 service in India. There are variations of access to the local loop depending on two factors (Sridhar, 11 June 2004):

• The element of the local loop offered as an unbundled element.
• The location chosen by the entrant to position its equipment.

Based on this, the unbundling can be divided into the following three categories:

1. Full unbundling
2. Shared Access
3. Bit-stream access

Full unbundling

Full unbundling is also called access to raw copper. In this scheme, the new entrant leases full access and exclusive use of the raw copper local loop. This is the copper connecting the customer's site to the local switch or to sub-loops connecting copper to a remote concentrator.

As a general proposition, there is a 'dedicated' copper pair from the serving local exchange to each customer premises. These copper pairs terminate at the local exchange end on a frame, called a Main Distribution Frame (MDF). When a new entrant orders an unbundled loop to serve a particular end-user, the incumbent will identify the twisted copper pair assigned to that customer's premises and disconnect it at the local exchange MDF. The individual copper pair will be cross-connected, either at the MDF or an intermediate frame, to transmission capacity, connecting the new entrant's transmission and multiplexing equipment collocated within the incumbent's local exchange. The new entrant will carry traffic over backhaul capacity between the incumbent's local exchange and the new entrant's switch, located remotely. The backhaul may be provided by the new entrant as a fibre or microwave link, or by the incumbent's leased capacity.

The new entrant can offer all kinds of services including voice. Full unbundling allows the new entrant to provide alternative local access service to customers previously connected to the incumbent. The incumbent maintains ownership of the copper pair, while control moves with the access seeker. This is depicted in the Exhibit below.

Shared access or line sharing

By the use of Digital Subscriber Loop (DSL) technology, it is possible to share the same loop for voice as well as for broadband connectivity. Line sharing involves

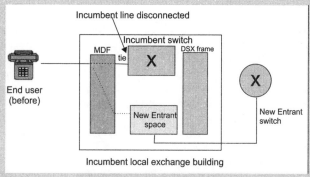

Source: www.itu.org

the incumbent using only the voice band frequency range of the physical copper to supply voice services to a particular customer. The frequency range outside the voice band can then be used by a competing service provider for the provision of high-speed data services. Under shared access, the incumbent and the new entrant share access to the high-frequency spectrum of the local loop. In this case the cooper pair will support both telephone service and broadband services provided through DSL technology.

End-users have the option of taking both Internet access and voice from the same carrier, or splitting their service by taking broadband services from one carrier (usually the new entrant) and voice services from another (usually the incumbent). A splitter is used to separate the telephone and data traffic. As a result, the customer gets voice from the incumbent and high-speed data services from a competing service provider.

Line sharing implementation options include the incumbent using its own equipment to hand over to the competing operator the high-frequency component of the copper for the provision of high-speed data services; or the competing operator taking physical possession of the copper and using its own equipment to hand back the low-frequency component to the incumbent for the provision of voice services.

The important issue in this approach is the need for collocation of the required equipment by the access seeker in the incumbent's exchange premises in or near the MDF room. There are a number of advantages accruing to customers, access seekers and the incumbents. The required investment for upgrading the local loop can be made by the access seekers to support the investments of the incumbent. Variety of technologies can be deployed by the new entrants to provide innovative services to the advantages of end-users.

Bit-stream Access

In this type of access, the incumbent allocates spectrum to a new entrant but maintains full control over the subscriber's line. New entrants can only supply services designated by the incumbent. Bit-stream access does not involve any physical access to copper pairs by the new entrant. The incumbent maintains control over the subscriber lines and provides equipment and modems for the new entrant. This method is also known as wholesale access.

Bit-stream access does not require collocation. The incumbent is responsible for creating the high-speed access link to the customers' premises and giving access seekers the upstream portion of the data/video communication. The access seekers are locked into whichever technologies are implemented by the incumbents. The responsibility for maintenance, fault repair, provision and other servicing lies completely with the incumbent. However, bit-stream access has an advantage in that the deployment process is simple and interconnection between access seekers and incumbents occurs at the data stream level, and not at the physical copper cable level.

and remote areas of the country. Sharing of the government operators' infrastructure in these areas can have a positive impact.

TRAI had suggested way back in April 2004, non-discriminatory Local Loop Unbundling (LLU) so that the access networks of the Basic Service Operators (BTOs) could be shared with Internet Service Providers (ISPs) and other competing operators for improving broadband penetration in the country. However, BSNL and MTNL view LLU as infringements on their property rights and hence have prevented the government from implementing the same. TRAI recommended LLU of only lines installed five years back. The newer lines are left to be utilized by the owners themselves. LLU will not be successful if competing operators could share only older lines with longer loop lengths and poor line conditions as these are not suitable for broadband connectivity.

However, case studies of UK and USA where local loop unbundling was allowed indicates that the high price of unbundling set by the incumbents, complicated terms and conditions negotiated by the incumbents and the general resistance of the incumbents to unbundle their local loop were the main causes of their poor success. Pricing of unbundled local loops as set by the regulator is crucial. If the price is set too high, it may not be attractive for competing operators. If it is too low, investment incentives are destroyed. Ideally prices should reflect the long run incremental cost plus a mark-up to ensure that the common line costs of the incumbent carrier can be recovered. Hence unless it is a win–win situation in terms of pricing and associated terms and conditions for both the access seekers and the incumbents, local loop unbundling would not be successful (Sridhar, 17 July 2006).

Technology Advances

Convergence in technologies has created challenges and opportunities for the BTOs in the country. The BTOs had to be agile in embracing these technology advances to be successful in the competitive marketplace. Most of the BTOs in developed markets have upgraded their networks and service to provide voice, data (Internet access) and video (Internet protocol television [IPTV]) which is referred to as 'triple-play' service. Add to that mobility, and the service can become 'quad-play'. Since voice has been commoditized and earns less revenue, the BTOs had to provide other value-added services to recover the high cost of fixed-line networks.

BROADBAND ACCESS SERVICE

The Broadband Policy announced by Government of India in 2004 defined the Broadband as

An 'always-on' data connection that is able to support interactive services including Internet access and has the capability of the minimum download speed of 256 kilobits per second (kbps) to an individual subscriber from the Point of Presence (PoP) of the service provider intending to provide Broadband service where multiple such individual Broadband connections are aggregated and the subscriber is able to access these interactive services including the Internet through this PoP. The interactive services will exclude any services for which a separate license is specifically required, for example, real-time voice transmission, except to the extent that it is presently permitted under ISP license with Internet telephony.

One of the technologies by which broadband service can be provided to end-consumers by the BTOs is DSL (see Box 2.4 for a primer on DSL).

Though 2007 was declared as the Broadband Year, there was not much progress in the deployment of broadband services. The reasons for the same are listed below:

1. Lack of enthusiasm in private operators to roll out services beyond high-revenue city subscribers as pointed out by Sridhar and Malik (2007).
2. Condition of the local access loops of BSNL, the government operator which prevented provisioning of DSL services.

Figure 2.5 shows the broadband subscriber base in India.

The TRAI, in its 2007 recommendations, listed the following requirements for the uptake of DSL broadband services in the country:

Box 2.4   Digital Subscriber Loop Technology

Telephone engineers initially developed the local loop of the PSTN to carry POTS voice communication and signalling: no requirement for data communication as we know it today existed. For reasons of economy, the phone system nominally passes audio between 300 and 3,400 Hz, which is regarded as the range required for human speech to be clearly intelligible. This is known as the voice band or commercial bandwidth. The local *telephone exchange* or *central office* generally digitizes speech signals into a 64-Kbit/s data stream in the form of an 8-bit signal using a sampling rate of 8,000 Hz, therefore, according to the Nyquist theorem, any signal above 4,000 Hz is not passed by the phone network (and has to be blocked by a filter to prevent aliasing effects).

The laws of physics, specifically the Shannon limit, cap the speed of data transmission. For a long time engineers believed it impossible to push a conventional phone line beyond low-speed limits (typically under 9,600 bit/s). In the 1950s ordinary twisted-pair telephone cables often carried 4-MHz television signals between studios, suggesting that the Shannon Limit would allow transmitting many megabits per second. However, these cables had other impairments besides Gaussian noise, preventing such rates from becoming practical in the field. The 1980s saw the development of techniques for broadband communications that allowed the limit to be greatly extended.

The local loop connecting the telephone exchange to most subscribers has the capability of carrying frequencies well beyond the 3.4 kHz upper limit of POTS. Depending on the length and quality of the loop, the upper limit can be tens of megahertz. DSL takes advantage of this unused bandwidth of the local loop by creating 4,312.5-Hz-wide channels starting between 10 and 100 kHz, depending on how the system is configured. Allocation of channels continues at higher and higher frequencies until new channels are deemed unusable. Each channel is evaluated for usability in much the same way as an analogue modem would be on a POTS connection. More usable channels equates to more available bandwidth, which is why distance and line quality are a factor (the higher frequencies used by DSL travel only short distances). The pool of usable channels is then split into two different freqency bands for upstream and downstream traffic, based on a preconfigured ratio. This segregation reduces interference. Once the channel groups have been established, the individual channels are bonded into a pair of virtual circuits, one in each direction. Like analogue modems, DSL transceivers constantly monitor the quality of each channel and will add or remove them from service depending on whether they are usable.

One of Lechleider's (Lechleider, 1991) contributions to DSL was his insight that an asymmetric arrangement offered more than double the bandwidth capacity of symmetric DSL. This allowed ISPs to offer efficient service to consumers, who benefited greatly from the ability to download large amounts of data but rarely needed to upload comparable amounts.

The following figure illustrates how asymmetric digital subscriber loop (ADSL) is deployed in the frequency space of the copper pair of the last mile connection.

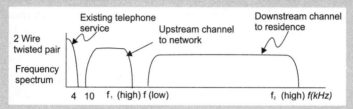

The commercial success of DSL and similar technologies largely reflects the advances made in electronics that, over the past few decades, have been getting faster and cheaper even while digging trenches in the ground for new cables (copper or fibre optic) remains expensive. One of the other major factors in the commercialization of DSL by BTOs in the 1990s, especially in the US was the competition in the local exchange carriage enforced by the Telecom Act of 1996.

The figure below gives a schematic diagram of a typical DSL deployment.

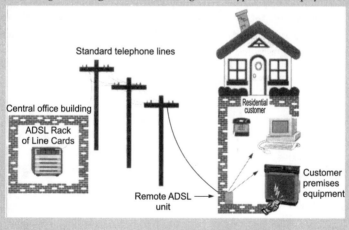

1. The government should increase the target fixed for BSNL and MTNL for provision of broadband connections during 2007–8. For this purpose BSNL and MTNL should be encouraged to appoint franchisees for providing broadband services to supplement their efforts. There should be total flexibility in developing a commercial model. Any procedural restrictions/limitations should be addressed immediately.

2. The government should encourage Indian manufacturers to produce more customer premise equipment (CPEs) used to provide broadband using DSL technology.
3. Standardization of DSL CPEs should be done by the Telecom Engineering Centre (TEC) immediately in a time-bound manner and information of all CPEs conforming to specifications should be displayed on the TEC website for the benefit of the customers.

Due to non-viability of DSL, especially in remote and rural areas of the country, the Indian government also auctioned spectrum for broadband wireless access which is discussed in a subsequent chapter.

WIRELESS NETWORK ACCESS

Wireless fidelity (Wi-Fi) was another access technology that competed heavily with fixed-line service, especially for broadband Internet connectivity. The Indian government had unlicensed the spectrum (2.4–2.4835 GHz) for IEEE 802.11 networks for outdoor usage (TRAI, June 2005). This enabled ISPs and other access service providers to offer Wi-Fi-based hotspot access at public spaces for Internet connectivity, thus bypassing the need for fixed line access. Realizing the need for embracing these new technologies, BTOs started launching Wi-Fi access service at a very low US$ 4 per month, targeted at college students, educational institutes and corporate users. The technology landscape continued to evolve, with wireless broadband technologies such as Wi-Max and Wi-Bro being adopted in developing countries such as India almost at

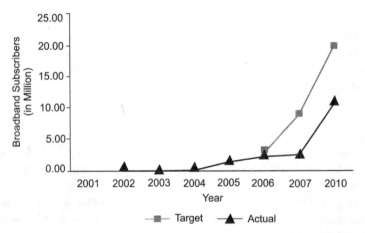

Figure 2.5   Broadband Subscriber Base in India

the same pace as developed countries. This posed a challenge for the BTOs to continuously embrace cutting-edge technologies to remain competitive in the marketplace.

The debate on the future of the landline industry started in the year 2003 itself (Sridhar, 15 November 2003). Box 2.5 gives some important tips for the survival of landline industry in India.

INTERNET PROTOCOL TELEVISION

Internet protocol television (IPTV) is a system where a digital television service is delivered by using Internet protocol over a network infrastructure, which may include delivery by a broadband connection. A general definition of IPTV is television content that, instead of being delivered through traditional broadcast and cable formats, is received by the viewer through the technologies used for computer networks. For residential users, IPTV is often provided in conjunction with video on demand and may be bundled with other Internet services. Internet protocol television is typically supplied by a BTO using a closed network infrastructure. In businesses, IPTV may be used to deliver television content over corporate Intranets. Internet protocol television provides opportunities for BTOs to improve the utilization of the local loop and earn much needed revenue. Realizing its potential the government BTOs started delivering IPTV services in select locations of the country in September 2007. However, the IPTV market is in the very early stages and regulatory guidelines are needed to address content ownership, cable TV rights and exclusivity, and tariff fixation.

A Cause-effect Model

Having analysed the various factors influencing the basic telecom services in the country, is it possible to predict the future of the market, subscriber base and the financials of the operators? The first such attempt is to build a causal loop diagram (refer to Figure 2.6) that shows the existence of all major cause-and-effect links, indicating the direction (cause -> effect) of each linkage relationship. A link is positive (or negative) if a change in the causal element produces a change in the same (or opposite) direction for the effect element. A closed sequence of causal links represents a causal loop. An even/odd number of negative polarity (that is, direction of change) links in a loop results in a positive/negative feedback.

To get an integrated view, we start with the main variable: subscribers. Households are subscribing to the service and getting converted into subscribers. Hence, an increase in the number of households in any circle/

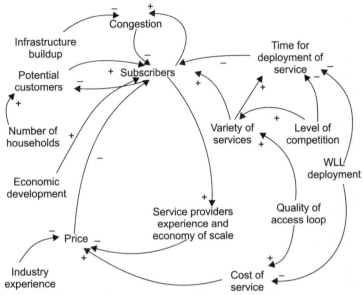

Figure 2.6 The Causal Loop Diagram of Basic Telecom Services

Box 2.5   Death Knell for Landlines?

While we witnessed the crawl of landline subscriber base in India from 12 million in 1995 to about 50 million currently, the cellular mobile subscriber base grew from zero to more than 750 million during the same period. Unified licensing further encourages the deployment of cellular like wireless local loop services even by the BTOs. So, the mute question is whether there is any future for landline service business in the country?

Typically landline services are considered as 'basic' services because of their limited ability to provide Plain Old Telephone Service (POTS). Until recently, only wired local loop (landline) from the telephone exchange was used by the BTOs to provide POTS. Traditionally 'teledensity', a measure of telecom penetration, is also measured using number of landlines deployed. Cost of installing a landline is still high in India. However, tariffs for basic landline service in India, is always kept low, often below cost, as it exists currently (Rs 0.40 per minute) so that it is affordable to many. Though lower tariffs are in favour of landlines, quick deployment, lower per line cost, competition and the value-added mobility feature have caused tremendous growth in mobile services, in our country. There are service areas where customers of government-owned BTOs such as MTNL and BSNL have started surrendering their landlines and switching over to mobile services. Given this scenario, should the BTOs, at all be investing in landline services? If so, what is their target market and how should they sell their services?

Betting on voice service alone is not sufficient for landline service providers. BTOs have to redefine their landline service not just as Plain Old Telephone Service (POTS) but as a value-added service capable of providing high-speed Internet connectivity. Since most of the BTOs have their Internet service provider (ISP) licenses, they should bundle ISP connectivity with their basic service. Broadband connectivity is the key to the survival of landline services, especially in metro areas. With the widespread deployment of DSL, BTOs are in a position to provide high speed Internet connectivity through the same landline infrastructure.

Landline service is a 'household' based service whereas mobile services are targeted at individuals and to some extent, personalized services. In smaller cities and towns, there are still households rather than individuals that purchase communication services. Apart from residential customers, targeting high-volume business customers is another option and is being pursued actively by private BTOs. However, business customers are more demanding than residential customers. Normally corporate customers buy telecom services in bulk and require one service provider to provide an integrated solution for catering to their voice, data and video communication needs. BTOs should equip themselves to provide one-stop integrated communication services rather than just operate as just another telephone company. For the corporate customer who buys telecom services in bulk, quality of service such as high availability, low fault repair time, and higher grade of service are more important than mere connectivity. Customer services such as customized billing. The government BTOs who control more than 90 per cent of the landline services market are notorious for their poor customer service. Customer Relationship Management (CRM) was never heard of in the basic services industry, thanks to lack of competition. Today, BTOs should actively adopt CRM much like their mobile counterparts, to acquire appropriate customers and retain their existing customer base. Marketing of telecom service is another area that was never done by the BTOs, thanks to BSNL and MTNL, the erstwhile monopolies. Today, BTOs have to resurrect their marketing efforts to build brands and capture loyalty of customers for their landline services.

Unless BTOs promote other value-added services, target households, and improve their marketing and customer service efforts, landline service may just not be able to sustain the onslaught of mobile services in India.

area will lead to an increase in the number of subscribers and will deplete the potential subscribers. The economic development of an area/circle increases the level of the telecom services. As the number of subscribers of a telecom service increases, the amount of information traffic and the frequency of requests for connections will increase. Thus the congestion faced by the subscribers of the service will increase, which will in turn deteriorate the network performance. However, if the service provider synchronizes the infrastructure build-up (that is, setting up of switches, access loops and trunks), with the pace of building up subscribers, congestion is reduced.

The quality of the access loop decides the type of services a service provider can provide and a subscriber can subscribe to. Economically, the cost of the access loop is one of the major components of the total infrastructure cost of telecom services. Hence, as the quality of the access loop improves, the variety of services being provided as well as the cost cf providing services increases. Increase in the cost of service will force the price of subscription to the service to increase, which in turn will show its negative effect on the subscriber levels based on their price elasticity of demand for services. The price for services decreases with an increase in the experience of the service providers and economies of scale of deployment. The variety of services is not only governed by the quality of the access loop but by the level of competition also. If there is no competition in the marketplace, then the monopolist will not feel motivated to provide higher quality services even though the access loop is capable of supporting such services. However, in order to provide multiple services, the service providers have to augment the quality of their access network. Pricing of the telecom services is the mechanism that provides the linkage between supply and demand, and it also sets the threshold level for being able to provide sustainable service. The operator has to make a trade-off between affordability and sustainability, that is, to reduce the cost of services to make it affordable for a wider segment of subscribers, and the need to generate enough revenue for sustainable infrastructure investment.

Figure 2.6 provides a synergistic view of the basis telecom services market. Techno-economic simulations of the model can be used to replicate the conditions of the telecom technology and the economic environment of the target market so that growth of telecom services can be investigated and monitored by researchers, planners and managers. Once calibrated, such models can be used to determine the effect of variation of model parameters on different outcome variables, especially the growth in subscribers for basic telecom services.

* * *

Though the landline was considered as a technology for providing universal service, it failed to meet the needs of developing countries such as India. Indications are that wireless technologies will pave the way for broadband penetration and ubiquitous connectivity. Table 2.10 summarizes the events in the telecom reforms of the country at various points:

Table 2.10   Sequence of Events in Basic Telecom Services

Year	Event
1985	Department of Telecommunications
1986	MTNL and VSNL
1989	Telecom Board/ Telecom Commission
1994	National Telecom Policy
1996	Basic telecom licence guidelines announced and licences issued in six circles for entry of one more operator
1997	TRAI
1999	New Telecom Policy; shift to revenue sharing; Universal Service Obligation Fund
2001	Second round of Basic Telecom Service licensing
2003	Unified Access Service Licence implemented; ADC implemented by TRAI
2004	TRAI recommendation on Local Loop Unbundling

3

Cellular Mobile Services
The Indian Success Story

Quick deployment, competition, advancement in technologies, and reduced cost of access have propelled the growth of mobile services in India much like in other developing countries. India currently boasts of the second largest subscriber base in the world even though it adopted second-generation digital cellular technology only in 1995. The number of mobile connections in India stands at about 700 million (at the end of September 2010) while the number of wire-line connections has been stagnant at about 40 million for the past few years. With the monthly addition of about 6–7 million subscribers, the target of 500 million by 2010 as set in NTP 1999 was exceeded well ahead of time. The widely used S-curve model of growth as illustrated by Singh (2008) was used to predict the mobile density and hence the subscriber base until the year 2015. The mobile subscriber base is expected to grow to 1 billion by 2014. Table 3.1 gives an overview of the important milestones in the Indian mobile services industry. This chapter enumerates the technologies, licensing process, and market structure of this industry.

Mobile Technologies

Over the past decade, wireless networks have made giant strides, moving rapidly from first-generation (1G) analogue, voice-only communications, to second-generation (2G) digital, voice and data communications, and further to third-generation (3G) wireless networks as a convergence of wireless and the Internet.

Table 3.1   Important Milestones in the Indian Mobile Sector

Year	Event
1991	DoT began the process of introducing private participation in the sector by inviting bids for two licences in each of the metros (Delhi, Mumbai, Kolkata, and Chennai)
31 July 1995	First cellular mobile service was started in Kolkata by Modi Telstra's MobileNet service. Bids were invited and two licenses were awarded in the remaining eighteen Licence Service Areas (LSAs) of the country using a single-stage sealed bid auction.
1999	Introduction of revenue share-based licence fee
2001	Government operators (BSNL and MTNL) were given the third mobile operator licence. The government issued guidelines for the fourth cellular licence. One licence in each LSA was awarded using multi-stage auction.
2003	The TRAI recommended that fixed-line service providers should be allowed to provide WLL-M service. In November 2003, UASL was introduced.
2003	The 'receiving party pays' scheme paved the way for CPP scheme.
2005	Foreign Direct Investment (FDI) limit applicable to the telecom sector raised from 49 per cent to 74 per cent.
2007	Government awarded new licenses on a fixed fee basis to a number of players. Additionally, under the 'crossover spectrum' policy, CDMA and GSM operators were made eligible to use both the technologies under the same licence.
2008	DoT announced third-generation (3G) guidelines.
2010	3G and broadband wireless access spectrum auction.
2010	Mobile Number Portability started at Rohtak, Haryana.

First-generation Wireless Network

The 1G networks were developed and installed in the early 1980s (Vriendt *et al.*, 2002). All the 1G systems used analogue technology that relied on Frequency Division Multiple Access (FDMA) methods to create multiple radio channels for multiple users. Analogue technology is an electronic transmission technique accomplished by adding signals of varying frequency or amplitude to carrier waves of a given frequency of alternating electromagnetic current (Rapport, 1996). It is usually represented as a series of sine waves because the modulation of the carrier wave is analogous to the fluctuations of the voice itself.

Figure 3.1 shows the generic transport architecture of a 1G cellular radio network, which includes mobile terminals (MT), base stations (BS)

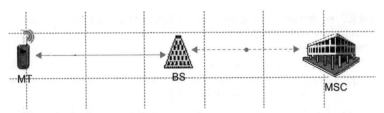

Figure 3.1   First-generation Wireless Network Architecture

and Mobile Switching Centres (MSC). The MSC maintains all mobile-related information and controls each mobile hand-off (Garg, 2001). The MSC also performs all of the network management functions, such as call handling and processing, billing and fraud detection (Rapport, 1996). The MSC is interconnected with the PSTN via trunks and tandem switches.

In the US, Advanced Mobile Phone System (AMPS) (Rapport, 1996) as the 1G wireless technology standard was released in 1983 using the 800–900-MHz frequency band with 30-kHz bandwidth for each channel with 666 channels (Garg, 2001). It was the first standardized cellular service in the world and was deployed apart from the US, in South America, China, and Australia. Total Access Communication System (TACS) is the European version of AMPS and was originally used in Britain in the 900-MHz frequency band. The TACS standard operates on the 900-MHz frequency band, allowing up to 1320 channels using 25-kHz channel spacing (Garg 2001). Finally, Nordic Mobile Telephony (NMT) is the classic cellular standard using 12.5-kHz channel spacing developed by Ericsson, the telecom equipment vendor, and was used in thirty countries around the world.

Second-generation Wireless Network

The second-generation (2G) standards were developed and installed in the early 1990s (Vriendt *et al.*, 2002). These systems digitize voice into discrete zeros and ones (also referred to as digitization), using either Time Division Multiple Access (TDMA) or FDMA or a combination thereof to create multiple access channels for subscribers.

Global system for mobile communication (GSM) (Rapport, 1996) is a globally accepted standard for 2G digital cellular communication. The GSM is the name of a standardization group established in 1982 to create a common European mobile telephone standard that would formulate specifications for a pan-European mobile cellular radio system operating at 900 MHz (Buchanan *et al.*, 1997). The GSM networks transmit data at 9.6 Kbps with a circuit-switched data transmission and allow up to eight users to share a single 200-kHz radio channel by allocating a unique time slot to each user (Garg, 2001). The GSM is used in the 900 and 1,800 MHz bands all over the world except for North America where the 1,900 MHz band is also used.

Enhancement of 2G GSM standard resulted in 2.5G technology, also referred to as General Packet Radio Service (GPRS). The GPRS permits packet-switched instead of circuit-switched data transmission at high-speed data connections of a theoretical maximum of 144 Kbps. The phase after GPRS is called Enhanced Data Rates for GSM Evolution (EDGE). The EDGE (Garg, 2001) is a radio-based high-speed mobile data standard that allows data transmission speeds of 384 Kbps to be achieved when all eight timeslots are used. The main idea behind EDGE is to squeeze out even higher data rates on the current 200 kHz GSM radio carrier, by changing the type of modulation used, whilst still working with the current circuit switches (Rapport, 1992).

High-speed Circuit-switched Data (HSCSD) (Carsello, 1997) is an enhancement of data services (Circuit-switched Data or CSD) of all current GSM networks. It allows one to access non-voice services at thrice the speed.

An alternative to the above technologies is the CDMA technology invented by Qualcomm in the US in the early 1990s, which has further improved system capacity and spectrum efficiency. The CDMA technology (Carsello, 1997) is a spread-spectrum technology that allows multiple frequencies to be used simultaneously. The CDMA technology codes every digital packet it sends with a unique key. The CDMA receiver responds only to that key and can pick out and demodulate the associated signal.

As seen in Figure 3.2, the 2G network architecture introduced new network architectures. First, the 2G system reduced the computational burden of the MSC and instead introduced the concept of 'Base Station Controller (BSC)' as an advanced call processing mechanism. The BSC is called a radio port control unit, which allows the data interface between the base station and MSC (Rapport, 2002) . Second, the 2G system uses digital voice coding and digital modulation. Finally, the 2G provides dedicated voice and signalling between MSCs, and between each MSC and PSTN. In contrast to the 1G system which was designed primarily for voice, the 2G has been specifically designed to provide data services as well (Garg, 2001).

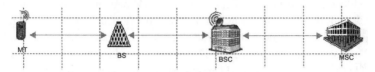

Figure 3.2   The Second-generation Wireless Network

Third-generation Wireless Network

In the early 1990s, the International Telecommunication Union (ITU) put forth a plan to harmonize the ongoing developments of a next-generation wireless network, referred to as the International Mobile Telecommunications (IMT)-2000. Under the IMT-2000, initiative, the 3G wireless system was split into two groups: the Universal Mobile Telecommunications System (UMTS) group (3rd Generation Partnership Project, 3GPP) and the CDMA 2000 group (3rd Generation Partnership Project 2, 3GPP2) (Rapport, 1996).

The UMTS was developed in 1996 with the sponsorship of the European Telecommunications Standards Institute (Vriendt *et al.*, 2002). In 1998, it was added to the IMT-2000 standards. It is also known as WCDMA because its infrastructure includes several WCDMA standards. The WCDMA technology (Carsello, 1997), which is an air interface standard in UMTS uses direct spread with a chip rate of 3.84 Mcps and a nominal bandwidth of 5 MHz. The UMTS is an upgrade of the GSM/GPRS that has enhanced its spectral efficiency up to six times.

The network architecture of UMTS is divided into the Radio Access Network (RAN) and the core network (Garg, 2001), as shown in Figure 3.3. The RAN contains the user equipment (UE), which includes the terminal equipment (TE) and mobile terminal (MT), and the UMTS terrestrial radio access network (UTRAN), which includes the Node-B and radio network controller (RNC) (Rapport, 1996). The core network (focused on packet domain) includes two network nodes: the serving GPRS support node (SGSN) and the gateway GPRS support node (GGSN). The SGSN monitors user location and performs security functions and access control. The GGSN contains routing information for packet-switching and provides interworking with external networks such as the packet data networks.

The WCDMA technology is network asynchronous, meaning that there is no synchronization between base stations. This implies that no additional source of synchronization is needed. In an asynchronous network, however, protocols must be carefully designed in order to

MT   node_B   MSC   SGSN   GGSN

Figure 3.3   The Third-generation Wireless Network (UMTS)

maintain successful handovers. A handover (or handoff) takes place when a mobile handset moves from one cell to another so that calls can be transferred to new channels without being interrupted.

The 3GPP standardization committee has enhanced the 3G technologies and has defined standards for High-speed Packet Access (HSPA) that uses Multiple Input Multiple Output (MIMO) to enhance the downlink data speed up to 42 Mbps (in a 5-MHz carrier). Efforts are being made to increase the speed up to 84 Mbps.

CDMA 2000 (Rapport, 1996) is another wireless standard designed to support 3G services as defined by the ITU and its IMT-2000. CDMA 2000 can support mobile data communications at much the same speeds as WCDMA technology (Garg, 2001). CDMA 2000 1X (IS-2000), also known as 1x and 1xRTT, is the core CDMA 2000 wireless air interface standard. The designation '1x', meaning *1 times Radio Transmission Technology*, indicates the same Radio Frequency (RF) bandwidth as IS-95: a duplex pair of 1.25-MHz radio channels.

CDMA 2000 1XEV-DO (Evolution Data Optimized) uses both CDMA and TDMA to maximize both individual user's throughput and the overall system throughput. It is standardized by 3GPP2 as part of the CDMA 2000 family of standards.

Fourth-generation Wireless Networks

Telecom vendors and service providers have started commercial deployment of a next-generation, truly broadband wireless cellular system, known as the fourth-generation (4G) (Dekleva *et al.*, 2007). The 4G system would allow for significantly higher bit rates per user (ranging 10–100Mbps), and would support the interoperability of diverse and heterogeneous wireless and mobile networks. This next generation of wireless technologies promises extensive opportunities for wireless services and applications, namely m-commerce and m-business. The factors that distinguish the 4G networks are roaming across networks, interoperability, and higher speeds. In 4G networks, the access to multiple wireless networks could also be facilitated by the use of an overlay network or by having intelligence in the networks. This would obviate the need for having multiple interfaces or adapters in user devices. A possible architecture of 4G networks is given in Figure 3.4.

One of the promising standards for 4G is Worldwide Interoperability for Microwave Access (WiMAX). This technology is being standardized by the ongoing work of Institute of Electrical and Electronic Engineers (IEEE) 802.16 Working Group under the name Wireless Metropolitan Area Network (WirelessMAN). WiMAX is promoted by the WiMAX Forum, an industry association promoting the technology and verifying its compliance with the IEEE standards. The core WiMAX standard was developed in 2001 and supported line-of-sight transmission in the 10–66 GHz frequency range. Amendment 802.16a supporting non-line-of-sight transmission in the range between 2–22 GHz bands was ratified in January 2003. In June 2004, amendment 802.16d—consolidating revisions 'b' and 'c' for quality of service, testing, and interoperability— was also ratified and is known as IEEE 802.16-2004. Amendment 802.16e-2005, which supports mobility, was concluded in 2005. The first products were certified in January 2006. Peek downstream data rates are anticipated at 12 Mbps and upstream 2–5 Mbps, but actual bit rates will most likely average 2–4Mbps. WiMAX functions on both unlicensed and licensed frequencies, but for industrial use the licensed spectrum will

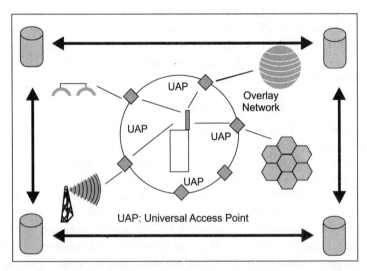

Figure 3.4   Possible Architecture of Fourth-generation Wireless Networks
Source: Dekleva *et al.*, 2007.

be used. The WiMAX forum is trying to focus development on the 2.4 GHz and 5.8 GHz bands of unlicensed frequency spectrum and the 2.5 GHz and 3.5 GHz bands of licensed spectrum. An effort is also being made to secure some spectrum below the 2 GHz band for WiMAX. Competing with WiMAX is Long-term Evolution (LTE) invented by the 3GPP committee. The GSM Association expects that LTE would take mobile broadband to the masses. With a peak downlink speed of more than 150 Mbps and a typical user downlink data rate of up to 50 Mbps, LTE promises to provide very high-speed broadband to end-users. Table 3.2 compares the key features of the different generations of technologies. Figures 3.5 and 3.6 explain the path of evolution of the above technologies.

Table 3.2   Comparison of Various Wireless Mobile Technologies

The Generation	Access Protocols	Key Features	Level of Evolution
IG	FDMA	Analogue, primarily voice, less secure, support for low bit rate data	Access to and roaming across single type of analogue wireless networks
2G and 2.5G	TDMA CDMA	Digital, more secure, voice and data	Access to and roaming across single type of digital wireless networks and access to 1G
3G and 3.5 G	CDMA 2000, W-CDMA, HSDMA, TD-SCDMA	Digital, multimedia, global roaming across a single type of wireless network (for example, cellular), limited IP interoperability, 144Kbps to several Mbps	Access to and roaming across digital multimedia wireless networks and access to 2G and 1G
4G	TBD	Global roaming across multiple wireless networks, 10Mbps, IP interoperability for seamless mobile Internet	Access to and roaming across diverse and heterogeneous mobile and wireless broadband networks and access to 3G, 2G, and 1G

Source: Dekleva *et al.*, 2007.

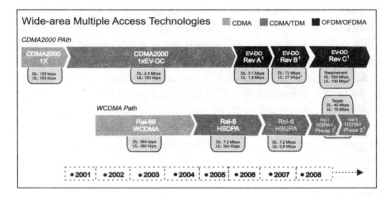

Figure 3.5 Evolution Path for 3G

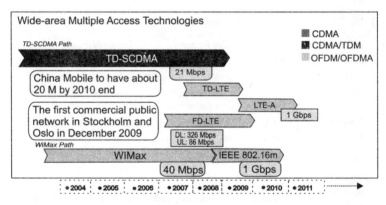

Figure 3.6 Evolution Path for 4G

It must be noted that Time Duplexing–Long-term Evolution (TD-LTE) has been jointly developed by China Mobile and Qualcomm, USA as a possible 4G migration path for all the different 3G technologies such as TD-SCDMA, CDMA2000 and WCDMA. Commercial deployment of TD-LTE started in 2010 and is set to pick up in the year 2011. Box 3.1 makes an argument for India to look at deploying 4G technologies (Pagare and Sridhar, 17 March 2010).

First Round of Licensing of Mobile Services in India

As in other countries, in India, the CMSPs are licensed to operate in designated geographical operating areas, referred to as 'circles' or LSAs. In India, there are twenty-three circles of which there are four metro

Box 3.1   India Should Leapfrog from 2G to 4G

In tune with the latest developments worldwide, the TRAI has issued a pre-consultation paper on 'Fourth-generation (4G) Mobile Wireless Broadband Services' earlier this month. While the 3G and even 2G spectrum allocation and licensing are in deadlock for the past one year due to interests of various concerned stakeholders, should we even think about 4G? Or, on the other hand, should we seriously consider leapfrogging from 2G to 4G as we did when we introduced 2G cellular mobile services in 1995, totally bypassing the first-generation analogue wireless technologies. Can we repeat a huge successful 2G story in 4G?

The 4G technologies are designed for services such as high-speed Internet connectivity, streaming multimedia services such as TV broadcasting, real-time high-resolution video conferencing, multimedia-based mobile commerce, to be delivered anytime anywhere using the ubiquitous IP. It is also completely interoperable across various types of 2G and 3G networks.

With the mobile subscriber base crossing 525 million, 12–14 operators vying for an average of 6–7 MHz of spectrum in each service area, landline deployment showing negative growth rate thus hurting broadband penetration in the country, is there a better reason for the Indian operators to leapfrog from the existing relatively inefficient 2G to efficient and futuristic 4G technologies and services? However, the moot question is whether the operators will get the desired radio frequency spectrum to offer 4G services, while the murky spectrum allocation for 3G is still going on.

The government earmarked 2.3 GHz band for broadband wireless access (BWA) with allocations up to 20 MHz for each operator in each service area. Though the BWA guideline is technology agnostic, the preferred and widely deployed technology in this band is WiMAX. The government is also trying to delink the spectrum allocation in 2.3 GHz from the 2.1 GHz allocation for 3G services, thus paving the way for quick deployment for broadband wireless access in the country.

However, another 4G technology known as LTE is also reaching maturity in commercial deployment. The first LTE network rollout was carried out by the Swedish operator TeliaSonera on 14 December 2009, with many operators in countries such as Japan, South Korea, and the US testing it out. There has been a strong movement in the US and Europe to deploy 4G technologies such as LTE by refarming the existing 2G spectrum bands. The spectral efficiency (that is, capacity per Hertz of spectrum per cell) is found to be almost forty times using 4G technologies compared to 2G. Moreover, unlike 3G which requires a minimum of 2×5 MHz carrier, 4G technologies such as LTE can operate using a minimum of 2×1.25 MHz carrier. This will allow the existing GSM operators to partly or wholly refarm their currently used 900/1800 MHz spectrum for 4G LTE services, starting with a small amount of spectrum, alongside their existing 2G services, adding more spectrum as users switch over. Further, deploying 4G-LTE technology at 900/1800 MHz will also bring additional cost and logistic benefits to the operators as they can deploy this at the existing GSM sites. The 4G technologies apart from providing high-speed data and video services also allow operators to pack more of packet-based voice calls per MHz with superior quality. This will even make the arguments we keep hearing about spectrum scarcity redundant, enabling the operators to tailor their network deployment strategies to fit their available spectrum resources. As an example, Hong Kong's SmarTone-Vodafone, is basing its LTE strategy on refarming its existing 900/1800 MHz spectrum, while walking out of a 2.6 GHz spectrum earlier this year on the grounds that the price was too high. Even the CDMA operators worldwide are eyeing LTE as a global standard, thus reducing the friction with their rival GSM counterparts. While one might argue that the 4G handsets are more expensive to be affordable to the Indian masses, we do have the scale advantage of bringing down the prices, as demonstrated with 2G and now 3G-capable handsets.

By specifying the types of technologies and services to be offered (that is, 2G, 3G or 4G) while allocating spectrum, the regulator and the government have created bottlenecks in the deployment of broadband wireless access in the country. The operators on the other hand due to their vested protective interests have also been demanding new spectrum blocks (for example 2.1 GHz for 3G) while postponing their investment decisions on future technologies. It is time that both the government and operators wake up to the reality of this 'leapfrogging' advantage India has in mobile services. Let us move towards the technology agnostic spectrum policy thus removing any mandate on specific technologies and services to be used in specific spectrum bands. The owners of the spectrum, be it network operators or mobile virtual network operators, choose which technology to use (2G, 3G or 4G) for offering the needed services to their target customers as per their business models. Let the market decide the technology landscape for mobile services in the country instead of regulation delaying the much needed techno-economic benefits.

areas (namely Chennai, Delhi, Kolkata and Mumbai). The other circles are normally defined by state boundaries, except in case of large states such as Uttar Pradesh (UP) (namely UP (East) and UP (West)). Some of the smaller states were combined together as a single circle (namely North East covering all north eastern states such as Arunachal Pradesh, Manipur, Meghalaya, Mizoram, Nagaland, and Tripura). These circles are categorized as A, B, and C. The categorization is based on the expected revenue potential with Category C circles at the lower end of the scale. Table 3.3 lists the categorization of circles in India for mobile services. Around 2005, the Chennai LSA was merged with the Tamil Nadu circle.

Table 3.3   Circles for Mobile Services

Metros	Category A	Category B	Category C
Delhi	Maharashtra	Kerala	Himachal Pradesh
Mumbai	Gujarat	Punjab	Bihar
Chennai	Andhra Pradesh	Haryana	Orissa
Kolkata	Karnataka	Uttar Pradesh (W)	Assam
	Tamil Nadu	Uttar Pradesh (E)	North East
		Rajasthan	Jammu and Kashmir
		Madhya Pradesh	
		West Bengal, Andaman and Nicobar*	

Note: * West Bengal and Andaman and Nicobar were separate circles initially and later were merged.

To start with, the Indian cellular market adopted a duopoly market with licenses given to two CMSPs. Global Systems for Mobile (GSM) was mandated as the technology to be adopted. The licensing process for cellular mobile services started in India in 1992, initially for the four metro areas of the country using a 'Beauty Parade' method. A schedule of minimum licence fee to be paid by the licensee for each of the metros was specified (see details in Table 3.4). From year four onwards, the operator was to pay either the annual specified fee or Rs 5,000 per subscriber, whichever was higher.

The bids were evaluated based on certain pre-qualification criteria including bidders' financial net worth, and prior experience in telecom service provisioning. The financial bids were evaluated based on the monthly rental fee to be charged by the bidders for the first three years. The operator was to pay either the annual licence fee as given in Table 3.4 or Rs 5,000 per subscriber, whichever was higher. Based on the rentals quoted by the successful bidders, the ceiling rental charge for cellular mobile service was fixed at a low Rs 156 per month. The DoT, the telecom policymaker and licensor in India, planned that about 60 per cent of the license fee would have to be financed by the higher call charges. Metro licences were issued and the first digital cellular service started in the metros in 1995.

In August 1995, licenses were awarded to two CMSPs in each of the other eighteen circles (currently nineteen). After the pre-qualification round, a single-stage auction procedure was used and the highest bidder was selected for the licence. The second highest bidder was asked to

Table 3.4   Licence Fees for Metros in the First Round of Licensing in 1994
(in Rs Crore)

	Mumbai	Delhi	Kolkata	Chennai
Year 1	30	20	15	10
Year 2	60	30	30	20
Year 3	120	80	60	40
Years 4–6	180	120	90	60
Year 7	240	160	120	80

Source: Desai, 2006.

match the winning bid. Initially the licence was given for a period of ten years. The rentals were fixed as the same for metros (for details regarding the licensing procedure, the reader is referred to Desai, 2006 and Jain, 2001).

The award of licenses through a single-stage auction represented maximum spectrum licence revenue for the government treasury and reduced the regulatory workload of allocating spectrum. As a mechanism for capturing the value of the spectrum, the single-stage auction in the first round of licensing was spectacularly successful. Since the telecom landscape was then evolving in India, the bidders had little idea how to evaluate market potential and bid accordingly. The bidders fell prey to the classic 'Winner's Curse' problem. The amount of licence fee bid for the eighteen circles amounted to a total of Rs 19,945 crores as given in Table 3.5. The licence fee payable over the licence period was two to three times the annual fixed service revenue of the incumbent monopoly government operator BSNL (ICRA, 2002). As can be seen in the Table, the bidding was higher in Category A circles and lower in Category C circles. The CMSPs were liable to pay their licence fee commitments over the licence period. The licence fee payments were roughly equal to annual revenues and imposed a heavy financial burden on the initial years of operation.

In 1999, the government realized that most of the operators were not able to pay even the annual fee towards the bid amount. The licence fee payable till 31 July 1999 was as high as twenty times the annual cellular services revenue for the financial year 1999. Study on the financial performance of CMSPs by the TRAI (TRAI, 1999) indicated that most of the operators accumulated losses till 1999. The accumulated losses were higher for Category A circle operators than for operators in Category B and C circles, mainly because of higher licence fee payments. Though the

Table 3.5   Licence Fee Commitments for Cellular Mobile Services

Circles (Category)	First Round (1995)			Second Round (2001)
	No. of Licences issued	Aggregate Licence Fee Bid (in Rs Crore)	Aggregate Licence Fee Paid After Revenue-sharing in 1999 (in Rs Crore)	Licence Fee (in Rs Crore)
Andhra Pradesh (A)	2	2,002	569.51	103.01
Gujarat (A)	2	3,588	1,020.88	109.01
Karnataka (A)	2	2,786	770.74	206.83
Maharashtra (A)	2	3,315	913.21	189.00
Tamil Nadu (A)	2	1,672	282.91	233.00
Haryana (B)	2	480	116.90	21.46
Kerala (B)	2	1,034	295.06	40.54
Madhya Pradesh (B)	2	10.20	29.12	17.45
Punjab (B)	2	2,532	849.00	151.75
Rajasthan (B)	2	764	217.31	32.25
Uttar Pradesh (W) (B)	2	422	115.90	30.55
Uttar Pradesh (E) (B)	1	812	138.26	45.25
West Bengal, Andaman and Nicobar (B)	1	42	12.24	10.00 (fourth licence issued in 2004)
Assam (C)	1	1.00	0.38	5.00 (fourth licence issued in 2004)
Bihar (C)	2	273	89.49	10.00
Himachal Pradesh (C)	2	30	8.54	1.10
North East (C)	2	4.00	2.42	2.00 (fourth licence issued in 2004)
Orissa (C)	2	178	58.48	5.00 (fourth licence issued in 2004)
Total		19,945	5,490	1,213

Source: ICRA, 2002.

operating revenue in metros was higher for the operators, all operators in
the metros also had accumulated losses till 1999.

To correct the above situation, the government intervened and proposed a migration package for the existing licensees from a fixed licence fee format to a revenue-sharing one. As per the package, the operators were given the option to pay an entry fee equivalent to 2.8–2.9 years of their committed licence fee plus an annual licence fee calculated as a percentage of their adjusted gross revenue (AGR). All the operators paid the arrears and migrated to a revenue-sharing scheme, except for one operator who could not pay the entry fee and subsequently lost its licence in UP(E), UP(W), Bihar and Orissa. Details of the licence fee finally paid by the operators for each circle is given in Table 3.5, which is about 25 per cent of the licence fee bid amount. The revenue-sharing percentage recommended was 17 per cent of the AGR for incumbent migrating CMSPs and included contribution to the USO Fund, research and development, administration and regulation expenses. Adjusted gross revenue was defined as the 'Gross Revenue' accruing to the licensees by way of operations of the cellular mobile service and also included revenue on account of value-added services, supplementary services, and sale of handsets plus revenue accruing through resellers, and franchisees. Over the years, the Indian government has reduced the annual licence fees as presented in Table 3.6.

Second Round of Licensing of Mobile Services in India

Subsequently, the third operator licence was awarded to government-owned operators MTNL for Mumbai and Delhi metros, and Bharat Sanchar Nigam Ltd (BSNL) for the rest of the circles in 2001. Though the TRAI recommended that the annual licence fee should be paid by the government operators as well, NTP 1999 stipulated that the licence fee would be reimbursed by the government. The licensing process for

Table 3.6   Annual Licence Fee in Various Circles

Circles	Annual Licence Fees (Percentage of AGR)		
	2000	2001	November, 2003
Metros	17	12	10
Category A	17	12	10
Category B	17	10	8
Category C	17*	8	6

Note: * Ten per cent of AGR for the Andaman and Nicobar Circle.

the fourth cellular licence was initiated in 2001. Bids for the fourth operator licenses were invited for a lump sum to be paid as entry fees. Subsequently, licenses were issued in August 2001, using a three-stage ascending auction procedure. As compared to the single-stage auction used in the first round of licensing, in multi-stage simultaneous auction all the licenses were put up for bid at the same time and the bidders had an opportunity to bid for as many licenses as they wanted and participate in successive rounds of bidding. Simultaneous multi-stage auction generates more information for the bidders concerning licence values and facilitates the award of the bid to the ones who value them the most. In general, multi-stage auctions lessen the risks of bidders winning more than they want and at a higher cost than they would have desired. The winning bid amount for the fourth cellular licenses is given in Table 3.5. The reduction in the bid amount compared to the first round is evident due to the following reasons:

1. The cellular mobile service market had grown and showed a certain maturity level and hence made it possible for the operators to bid realistically considering the market potential.
2. The three-stage auction procedure allowed the bidders to revise their bids at each stage, thus reducing the probability of the 'winners curse'.

Though the government removed the requirement of GSM as the technology to be adopted in the fourth cellular licence guidelines, all the licence winners adopted GSM to make their network interoperable with existing networks. As can be noted in Table 3.5 most of the Category C circles were not picked up by any operator in 2001 and were picked up later.

Evolution of Unified Access Service Licence

During this period, the government also liberalized the BTS market which typically provided the traditional landline-based POTS. Jain and Sridhar (2003) provide details of the BTS operations and its growth in India. In the year 2000, BTS operators approached the government with a proposal that they could provide local access loop at a much lower cost using CDMA wireless technology. The BTS operators argued that quick deployment of wireless CDMA service provided high spectral efficiency and lower per line cost compared to landline services, and hence was definitely a better alternative compared to wired access loop for certain areas of the country (the reader is referred to Stallings, 2004 for details of GSM and CDMA technologies). This proved to be a direct competition for GSM-based CMSPs. The matter was referred to the ministerial Group on Telecom and IT (GoT-IT). In its verdict on 27 April 2001, GoT-IT noted that under NTP 1999 fixed service providers could provide all types of fixed services including wireless in local loop services. It also observed that if technology allowed limited mobility as an extended service to the advantage of the customers then it should be permitted. The government endorsed the recommendations of the GoT-IT committee (for details on the GOT-IT committee recommendations, the reader is referred to McDowell and Lee, 2003). The following clause regarding WLL-LM was specified in the basic service licence.

Basic service operator shall be allowed to provide mobility to its subscribers with wireless access systems limited within the local area (that is, the SDCA) in which the subscriber is registered. The system shall also be engineered so as to ensure that handover of subscriber does not take place from one SDCA to another SDCA while communicating.

The above move increased competition in mobile services. The majority of the calls made by subscribers (about 80 per cent) are intra-SDCA calls and since these calls were offered at low landline rates, the subscriber traffic shifted from existing CMSP networks to WLL-LM networks. After a couple of years of litigations between the BTS operators and CMSPs, finally the Indian government announced the UASL in November 2003 that allowed migration of basic service licence holders to provide full mobility-based services with a stipulated entry fee. This introduced CDMA-based mobile services in the Indian market. Under the UASL, the licensees could provide any access service using any technology (for details on UASL, refer to Prasad and Sridhar, 2008). The entry fee for UASL was equated to the price paid by the fourth cellular mobile licensee (about Rs 1,620 crores for a pan-India licence) in the respective LSA. Under the UASL, operators provide cellular mobile service using either GSM or CDMA. Licensees who provide services using either of these two technologies could pay the fixed fee again to become eligible for offering dual technology services in the respective LSA.

The UASL, though, had rollout obligation clauses much like the cellular mobile licence, altogether eliminated rural rollout obligation as required under BTS licence. Following are the rollout obligation clauses in the UASL (DoT, 2005):

1. At least 10 per cent of the District Headquarters (DHQs) will be covered in the first year and 50 per cent of the DHQs will be covered within three years of the effective date of licence.
2. The choice of DHQs/towns to be covered and further expansion beyond 50 per cent DHQs /towns shall lie with the Licensees depending on their business decision.
3. There is no requirement of mandatory coverage of rural areas.
4. On completion of one year from the effective date of licence and meeting the coverage criteria stipulated for the first year, the Performance Bank Guarantee (PBG) shall be reduced to Rs 10/5/1 crore(s) for category 'A'/'B'/'C' service areas on self-certification provided by the licensee. Further, on fulfilling the rollout obligations as stipulated in the licence agreement, the balance PBG shall be released on receipt of test certificate/test certificates issued by TEC in respect of coverage.

Competition in the Market

Until 2007, India had about seven to eight operators per circle. In early 2008 another seven to eight licenses on an average per LSA were given out, doubling the number of operators in each LSA. The stated objective of the government in giving out so many licenses was increasing competition in the market, and promoting the diffusion of mobile telephony to rural areas. While India has an average of fifteen operators per circle, countries other than India have an average of only four licence holders. A look at the 2G licenses issued in Europe indicates that the maximum number of licenses issued in any country is five (in Ukraine)

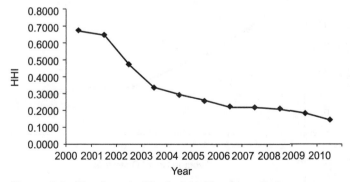

Figure 3.6   Trends in the Herfindahl-Hirschman Index

(Whalley and Curwan, 2006). Further, all the countries added only one or two licenses when introducing 3G technology.

One of the widely used measures of the competitiveness of the market is the Herfindahl-Hirschman Index (HHI). The HHI is a measure of the size of firms in relationship to the industry and an indicator of the amount of competition amongst them. It is defined as the sum of the squares of the market shares of each individual firm. The HHI can range from 0 to 1 moving from a very large number of very small firms to a single monopolistic producer. Decreases in the HHI generally indicate a loss of pricing power and an increase in competition, whereas increases imply the opposite. The HHI in India was around 0.15 at the end of December 2010. Figure 3.6 illustrates the decline in HHI over a time period.

However, the rate of decrease of HHI (that is, rate of increase of competitiveness of the market) levels off after four to five operators as indicated in Figure 3.7. This indicates that further entry of operators beyond four or five does not significantly increase the competitiveness of the market.

Price of Mobile Services

The DoT had fixed the maximum rental at Rs 156 a month based on the bids received for metros during the first stage of licensing. The DoT also placed a ceiling on call charges—a standard rate of Rs 8.40 per call minute, peak rate as high as twice the standard rate and off-peak rate as high as half the standard rate. However, realizing that the high call

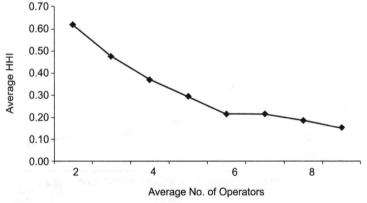

Figure 3.7   Number of Operators versus Competition Index

charges discouraged the callers, the TRAI raised the monthly rate from Rs 156 to Rs 600 per month in May, 1999. The standard peak-time call charge was also reduced from Rs 16.80 per minute to Rs 6 per minute. The pulse rate was reduced to 20 seconds, thus reducing the minimum call charges to Rs 2. The introduction of the third and fourth CMSPs increased competition leading to a drop in prices.

In 2002, the TRAI again reviewed and found that the average rental offered by the service provider in the market was Rs 195. The average call charges were Rs 2.03. Both were about a third of the ceilings that the TRAI had set (Desai, 2006). Following the review, further reduction of ceiling rates were specified by the TRAI in 2002. The metro rentals were reduced to Rs 475 and in other circles to Rs 500. The standard rate was reduced to Rs 4 per minute in metros (Rs 4.50 for other circles) and Rs 2 (Rs 2.25 for other circles) for each subsequent minute. However, the concession rates were subject to forbearance.

The introduction of WLL-LM in 2002 by the then BTS operators increased the amount of competition. The TRAI announced the CPP regime in January 2003, which fuelled a drop in prices. Until then the Receiving Party Pays scheme was in existence wherein the receiver of a mobile call also shared part of the call charges.

On 1 May 2003, the Telecommunication IUC regulation notified by the TRAI came into existence (Sridhar, 19 May 2003). As per IUC, mobile callers had to pay additional ADC to wire-line BTS operators towards maintenance of rural phone lines. This increased the per minute call charges. The TRAI has been reducing the ADC over the years and scrapped it all together from 1 April 2008. The reduction in effective mobile call charges due to the above regulatory interventions is shown in Figure 3.8. Increase in mobile subscribers can be attributed partially to the above drop in prices.

Mukherji (2008) argues that the political economy of three decisions, namely (i) phasing out of ADC (ii) implementation of UASL and (iii) raising the FDI from 49 per cent to 74 per cent in February 2005, rendered the regulatory and policy environment favourable for the foreign capital-dependent GSM operators.

Apart from rental and usage charges handset costs play a significant role in the penetration of mobile services, especially in developing countries such as India. In India about 60 per cent of the handset market is contributed by entry-level very low end (VLE) handsets. In 2006, the average price of a handset in India was about Rs 3,800 (Singh, 28 November 2006). However, realizing the market potential for the

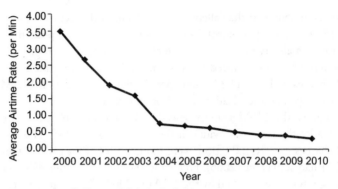

Figure 3.8   Effective Mobile Call Charges

VLE segment, companies such as Motorola, Nokia, LG, and Samsung aggressively introduced entry-level handsets for the VLE segment. Multinationals such as Nokia as well as home-grown vendors such as Micromax have cut down the price of their entry level handsets to about Rs 1,000. With Motorola and Nokia setting up manufacturing units in India, especially for catering to the handset market of the Indian subcontinent, the price of handsets is expected to reduce further, thus having a positive impact on mobile penetration in the country. The one-paisa revolution that created value for one paisa as mobile talk time is illustrated in Box 3.2 (Sridhar and Sridhar, 4 August 2010).

Revenue of Mobile Operators

India's mobile services are one of the cheapest in the world. Though low prices have fuelled demand, the Average Revenue per User (ARPU) per month continues to be well below the world average. The reduction of ARPU over a time period is shown in Figure 3.9.

The decrease in ARPU along with competition makes it financially difficult for the operators.

Taxes and Levies

Taxes and levies on mobile service providers are very high in India. There is the annual licence fee of 6–10 per cent levied on the AGR of the operator; an additional spectrum charge of 2–6 per cent as per the spectrum allotted; and ADC at 1.5 per cent of AGR. Apart from these, telecom subscribers pay service tax and education cess. There is also lack

Box 3.2   The One-paisa Revolution

The one paisa has lost its value. This may be evident in the fact that single-paisa coins are not available in India anymore. Cost-benefit considerations led to a gradual discontinuance of 1, 2, and 3 paisa coins in the 1970s. Thus coins are now available only in denominations of 10 paise, 20 paise, 25 paise, 50 paise, 1 rupee, 2 rupees, and 5 rupees.

However, the telecom revolution has created value for every single paisa for the common man. Though the battle of words and actions exists amongst the mobile operators in the country, thanks to intense competition, the telecom companies have revived the value of the Indian paisa. The only thing now that the paisa can buy today is phone talk time. For just one paisa, one can talk to someone in the farthest corner of India for one second, or send a short message service (SMS) of 160 characters to any one of the 600 million mobile phone users in India. In addition, one can extract more from the service provider if the bill plan is well chosen. Furthermore, you can even get paid one paisa for an SMS you send to your friends, which can be used as referrals by mobile advertising companies for their promotion campaigns. We call this phenomenon the 'one-paisa revolution'.

First we present evidence that the purchasing power of the rupee (or the paisa, as we argue here) has eroded considerably, taking into account inflation. The *Economic Survey* for 2009–10 notes that inflation year-on-year, measured in terms of the wholesale price index (WPI) exhibited significant volatility during 2008–9 with the volatility continuing during the current financial year. The volatility observed in the first half of 2008–9 was due to increasing international fuel and commodity prices which pushed WPI inflation to a high of 12.8 per cent. Apprehensions of shortages in agricultural production due to a deficient southwest monsoon this year are mainly responsible for increasing inflation. Average food inflation which was 7.56 per cent during fiscal 2008–9 increased to 13.54 per cent in the period April to December 2009. Overall food inflation in December 2009 was 19.77 per cent. Given this inflation, it is easy to believe that a single paisa does not have much value even for those at the lowest quintiles of the income distribution.

Now, if we agree that the paisa has lost the value of what it could potentially buy, with the exception of talk time, we need to understand the characteristics of a revolution to enable us to call the positive changes brought about by telecom aptly as a one-paisa revolution.

Revolution represents radical change. Even in its astronomical application, the word is used to describe change in position. While all revolutions involve change, not all change can be described as revolution. Change that may be called a revolution must have characteristics which a mere change lacks. That which is particular to the class of change which warrants the different name *revolution* is, either the *great speed*, the *great extent* of change, the *great number of people* affected and/or the *great degree* to which people and their lives are affected by change.

In India, the telecom revolution has taken place since 1995 primarily due to removal of policy and regulatory shackles. During this period, the mobile subscriber base in India grew from nothing to more than 600 million, connecting the largest number of people conceivably possible anywhere on the globe, with the exception of China. The resultant network effect has increased the demand for services, thus making it attractive for operators, equipment makers and content providers alike.

Further, there is also some evidence that the phone is powerful enough to replace face-to-face interaction in several instances. With every decrement in mobile airtime rates, the usage has been increasing. The minutes of usage per month is considerably higher than that witnessed in most of the countries. There is also reason to believe that the marginal utility of making a mobile phone call for people at the bottom of the pyramid is much more than for richer subscribers. The dual causality between economic development and mobile density is one of the reasons for the rapid increase in the rural subscriber base in the country. With people in all strata of the society, ranging from the fishermen of Kerala to women self-help groups in the hinterlands of Tamil Nadu getting notable benefits from the use of mobile services, there is support to believe that the changes brought about by telecom represent a revolution.

Another important concept that is making this one-paisa revolution possible is the 'price point pack' (PPP), an innovative marketing and packaging concept that is designed to sell a small portion or a single-use portion of the product at an affordable price. The first PPP ever introduced was some sixty years ago! Brooke Bond and then Lipton, the leading tea companies in India launched what was called the 'paisa pack' which was a paper envelope that contained tea worth one paisa (then 1/64th of the Indian rupee) which could make one or two cups of the beverage. Followed diligently by paan masala and then later by the Fast-moving Consumer Goods (FMCG) companies to sell shampoos in sachets, PPP has been successfully adopted by the Indian telecom operators to sell the perishable commodity often referred to as 'Erlangs' of network capacity.

Thus the one-paisa revolution has made growth more inclusive. However, the telecom industry in India continues to be much dogged by regulatory and policy uncertainties. At the minimum, subscribers and operators who have contributed to this revolution, expect that that the regulator and policymakers rise to the occasion and formulate appropriate and predictable policies for the country's development in an inclusive manner.

of transparency in the use of these levies for telecom-related projects. For example, the USL of 5 per cent is contributed by all telecom companies to the USO Fund. The USO fund was set up in April 2002 to provide financial support to BTOs, especially the government-owned BSNL for providing rural area connectivity.

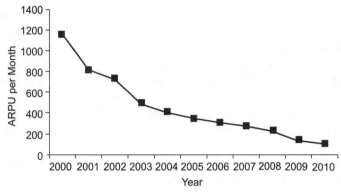

Figure 3.9   Trends in Average Revenue per User

In total, mobile operators in India pay about 30–5 per cent as taxes. The TRAI had pointed out this anomaly in the tax structure in one of its study papers (TRAI, 2005b), and compared it with other countries such as Pakistan, Sri Lanka, and China, where the tax burden is in single digit. Higher tax burden increases the annual expenditure and hence acts as a deterrent for further investment.

Interconnection Issues

In a multi-operator environment, network interconnection assumes great importance. The IUC is the charge payable by a service provider to another service provider or providers for the usage of the network elements for transit or termination of the calls. The Reference Interconnect Order of the TRAI in July 2002 specified interconnection charges to be mutually agreed upon between network operators. However, most of the operators could not come to mutual agreements on network interconnection charges (CRISINFAC, 2006).

Given this, the need was felt for an interconnection regulation that facilitated a level playing field for all operators, provided equitable revenue-sharing terms, and allowed open interconnectivity, so that subscribers could make any type of inter-network calls. Hence, to address these issues and facilitate equitable interconnection terms, in January 2003 the TRAI issued the interconnection regulation with respect to cost-based IUCs for inter-network calls. On 1 May 2003, the Telecommunication Interconnection Usage Charges regulation notified by the TRAI came in to existence (TRAI, 2003). The IUCs consist of (i) origination charges—the fees paid by the service provider for originating the call (ii) carriage charges—the charges for carrying the call across the circles/countries by the national long-distance/international long-distance operators (iii) termination charges—the fees charged by the network operator completing the call; and (iv) Access deficit charges—the deficit between calculated cost-based rentals and applicable rentals to be paid to the BTOs. The origination charges were made on forbearance. The originating service providers would retain the origination charges from the residual after payment of carriage, termination and access deficit charges.

Call termination assumes importance in a competitive market with a number of access providers (Sridhar, 27 March 2006). Call termination externality relates to termination of calls which originate on different networks. The effect arises because the person originating the call is not the customer of the operator who terminates the call. The terminating operator will be able to raise the price of termination or deny the termination with no direct effect on its own subscribers. The call termination externality represents a major barrier to effective competition and even in a competitive market may allow excessive pricing and poor quality of service. One way to circumvent this failure is through regulatory intervention.

In general, calls terminated within the caller's networks (referred to as on-net calls) are often charged less than those terminated in other operator's networks (that is, off-net calls). One reason for this is the termination charge (20 paise/minute as on December 2010) to be paid to the receiver's network for off-net calls. This fixed termination charge if set high will increase the user charges and hence will be detrimental to subscribers for all off-net calls. High termination charges will also lead to originating service providers' reluctance to interconnect with other operators. Moreover, higher termination charges also are likely to increase the difference between the prices of on-net and off-net calls leading to discrimination of off-net call users. If set low, it will lead to operators resorting to practices of denial of call terminations originating from other networks. Mobile service providers argue that the termination charges in India are one of the lowest in the world. Even the TRAI noted that mobile termination charges are twelve to fourteen times less than those prevailing in other countries (TRAI, 2006a). This leads to possible refusal of termination of calls or inadequate provisioning of interconnections with other networks for termination facilities.

Quality of Service

In a multi-operator environment, the quality of service (QoS) is an important factor determining the overall competitive performance of the service providers. The QoS features such as voice quality, service coverage, call completion rate, service access delay and Point of Interconnection (PoI) congestion depend on the quality and amount of telecom infrastructure deployed by the service providers. The TRAI issued the regulations for both basic and mobile services first in July 2000 which defined the QoS parameters and their benchmarks. Each service provider has to meet certain benchmark quality of standards as defined in the regulation. The first survey was conducted by the TRAI in 2001 (TRAI, 2001). These regulations were reviewed and revised later in July 2005 in consultation with the service providers. Revisions incorporated PoI congestion on outgoing calls as one of the parameters (TRAI, 2005e).

There are two parts to the QoS assessment. The objective assessment included audit of sample exchanges (including customer care centre), helpline and audit of sampled message switching centres (including customer care centre), and PoI congestion. The subjective assessment involved a survey of customer's satisfaction level with the CMSPs and unified access service providers (AUSPs). Details of the QoS parameters are given in Table 3.7.

The TRAI in its QoS study in 2006 (TRAI, 2006c) pointed out that on an average 59.52 per cent of the operators did not meet the subjective customer satisfaction benchmark levels. In the objective QoS assessment, one of the parameters in which almost all the mobile operators showed deterioration was the PoI congestion. On an average, only 32.84 per cent of the operators met the PoI benchmark levels. In as many as 398 PoIs, the congestion was alarming and well above the benchmark level of 0.5 per cent. At 249 locations, the PoI congestion was as high as 5 per cent (Sridhar, 13 March 2006). This high level of PoI congestion is due to inadequate junctions between the two networks that interconnect which lead to frequent blocking of calls and degraded quality of service. A closer look at the report indicates that the PoI congestion is mostly between private mobile operators and the government operator networks. This led to TRAI issuing directions to mobile service providers in various circles to improve network interconnections with other networks, especially the incumbent government operators' fixed-line networks. The QoS is still a

Table 3.7 Quality of Service Benchmarks for Mobile Services

Objective QoS Assessment		Subjective Customer Satisfaction Survey	
Parameters	Benchmark Level	Parameters	Benchmark Level
Accumulation of Down Time of Community Isolation	<24 hours	Satisfied with Provision of Services	>95%
		Prepaid Customers	>90%
		Post-paid Customers	>90%
Call Setup Success Rate	>95%	Per cent satisfied with help services	>90
Service Access Delay	9–20 sec	Per cent satisfied with network performance	>95
Standalone dedicated control channel (SDCCH)/ Paging Congestion	<1%	Per cent satisfied with maintainability	>95
Traffic channel (TCH) Congestion	<2%	Overall Customer Satisfaction	>95%
Call Drop Rate	<3%	Satisfaction with Supplementary Services	>95%
Per cent Connections with Good Voice Quality	>95		
Point of interconnect (PoI) Congestion			
Billing Complaints per 100 bills issued	<0.1%		
Per cent of Billing Complaints Resolved within four weeks	100		
Period of Refunds/Payments Due to Customers from the date of Resolution	100%		

Source: TRAI, 2006c.

concern and almost all the mobile operators are still struggling to meet the benchmark standards.

Mobile Number Portability

Number portability allows a subscriber to retain the same number if the subscriber switches to a different network, location or service. The

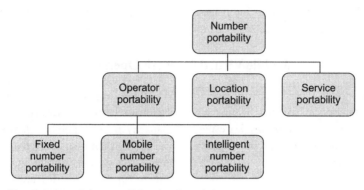

Figure 3.10   Schema of Number Portability

generic classifications of the number portability scheme are defined in Figure 3.10.

Mobile number portability (MNP) allows the subscribers to retain their existing mobile number when they switch from one service provider to another or from one technology (GSM/CDMA) to another of the same service provider. Portability benefits subscribers and increases the level of competition between service providers, rewarding those operators having better customer service, network coverage, and service quality. Keeping in view the growth of telecom services in India, and the level of competition the TRAI released its recommendations for MNP in 2006 so as to further enhance competition among service providers in the mobile sector and also to improve the quality of service and satisfaction of the mobile subscribers (TRAI, March 2006).

As per the recommendations, MNP was initiated by the Indian government through a neutral third party who shall establish a logically centralized database for implementing MNP (DoT, 2008). For the purpose of MNP, the telecom service areas of the country were divided in to two zones with eleven service areas each, with two metros in each area. Two third-party agencies were selected to provide the MNP service in the two zones. As per the MNP rules, in a multi-operator scenario when N operators are involved in the complete call setup considering 1 being the originating operator and N being terminating operator, (N-1) th operator shall be responsible for routing the call to the ported numbers. The up-gradation cost of their network will be borne by the operators themselves. The recipient operator shall be permitted to charge a fee (Rs 18 to start with) for successful porting, directly to the subscribers. The subscriber should wait for at least three months before the number can be reported to another service provider.

Mobile Virtual Network Operators

A mobile virtual network operator (MVNO) offers mobile voice and data services without owning any spectrum or network infrastructure. The MVNOs typically lease network capacity from a mobile network operator (MNO) and provide retail services using their own brand name, complementing the network with their own assets such as a strong brand, a loyal customer base, exclusive content, or an extensive distribution channel. The evolution of telecom markets in the US, Western Europe and parts of Asia have demonstrated the viability of the MVNO model, with MVNOs capturing 10–20 per cent market share in many markets.

Different Mobile Virtual Network Operator Models

An MVNO can adopt a range of business models to 'go to market'. At one extreme, it could act as a 'pure reseller' wherein it re-brands an MNO's service using its own brand name and sells through its distribution channels. On the other hand, it could adopt the role of a 'pure MVNO' in which the MVNO either buys or partners with third parties to provide all elements of the MVNO value chain beyond the spectrum and network infrastructure. The decision to adopt a given business model is governed by several factors, including the targeted scale of the business, level of in-house telecom expertise, extent of initial investment that the MVNO is willing to make, and the level of risk the MVNO is willing to undertake. The early MVNOs such as Virgin Mobile and Qwest in the US had to build their own back-office processes and platforms to complement an MNO's network. They accomplished this either by purchasing platforms and operating them in-house or through dedicated partnerships. However, with the increasing number of MVNOs entering the market a number of third parties emerged who could provide relevant processes Business Support Systems (BSS) and Operational Support Systems (OSS). These service providers are referred to as mobile virtual network enablers (MVNEs) (Mathew *et al.*, 2006).

Regulatory Aspects of Mobile Virtual Network Operators

Regulatory policies can alter the timing, scale and profitability of MVNOs. Regulators across the world have adopted varying positions with respect to regulating MVNOs. In Hong Kong, the regulator, the Office of Telecoms Authority (OFTA), used the 3G licensing timeframe to introduce MVNO regulations. It foresaw the opportunity to offer multiple services on the back of the same infrastructure and at the same time took precautionary measures to pre-empt a possible monopoly in the network operation business. The OFTA strived to ensure isolation of network operators from the business of service provision so as to afford maximum benefit to the end-consumer. This was achieved by using various levers within its power, including separate licenses for the two types of players, mandating a 40 per cent minimum as the capacity that has to be leased to non-affiliated MVNOs, and a non-discriminatory wholesale pricing regime. At the other extreme, Italy has a telecom market which has penetration of more than 100 per cent and yet the regulator did not permit MNOs to host MVNOs until 2011. This clause was included as part of the 3G licence agreements in order to make it commercially viable and to offset the large costs incurred by network operators in licence fees and telecom equipment. Such an approach could ensure adequate investments in new 3G infrastructure but could have a detrimental effect on customer choice and pricing. The facilitative approach followed by most European regulators and the Australian regulatory authority has resulted in a thriving MVNO industry in these countries piggybacking on multiple MNOs. Though they stopped short of mandating open access, regulators ensured that any incumbent that achieved 'Significant Market Power' status would open its network to MVNOs. This proved to be a safety valve against the creation of a monopoly, and resulted in MNOs actively scouting for partners with significant brand power or access to premium content.

The TRAI initiated the consultation process on MVNOs in 2008 after the controversial entry of UK-based Virgin Mobile into the Indian mobile markets in partnership with Tata Teleservices. Virgin Mobile sold its own subscriber identity module (SIM) cards and provided services packages targeted at youth. It adopted a pure reseller model, though at that time there was no mention of MVNOs in any of the telecom policy and guidelines documents. After getting stakeholders' views, the TRAI gave its recommendations at the end of 2008, with some clarifications on certain clauses in March 2009.

The salient features of the TRAI's recommendations are listed below:

1. The MVNO in a service area is an entity that does not have spectrum of its own for access services but is licensed to provide access services

to its customers through an agreement with any licensed access provider(s). The MVNO should not possess spectrum for access services in any manner including sharing and licensing of spectrum.

2. The MVNO is free to choose one of the following commercial models with an MNO in an LSA and the mutual agreement be submitted to DoT before commencing operations:

   a. *Full MVNOs*, which have their own core network including MSC;

   b. *Intermediate MVNOs*, which acquire a switched service, but either provide their own home location register (HLR) or share a jointly owned HLR with an MNO; and

   c. *Thin MVNOs*, which may provide access services with additional applications and content and they are not much different from pure resellers. These thin MVNOs are also called Enhanced Service Providers.

3. The MVNO shall be treated as a distinct service provider with its own licensing and regulatory framework, the essentials of which are given below:

   a. Entry fee for the MVNOs is as follows:
   Rupees 1 crore for metros/Category A LSAs; Rs 0.50 crore for Category B LSAs and Rs 0.25 crore for Category C LSA.

   b. The LSA of the MVNO shall be the same as that of the MNO. The MVNO shall partner with utmost one MNO in a service area. However, the MVNO could offer service anywhere within the LSA of the parent MNO as specified in the mutual agreement between the MNO and MVNO. Separate licence for each service area shall be required.

   c. The MVNO shall pay annual licence fee on AGR at the same rate as paid by the parent MNO in the service area of operation.

4. An MVNO shall be parented to only one MNO in a service area. The revenue from the MVNO shall be taken into account in the AGR calculations of the MNO and is liable for spectrum charges and licence fee calculations.

5. In a service area, merger of an MVNO with the parent MNO or another MVNO parented to the same MNO may be permitted. In case of merger of an MVNO with an MNO, the MVNO licence shall stand terminated.

The DoT was in concurrence with the TRAI on most aspects of the recommendations. There was some disagreement on points such as whether the MVNO should parent only one or two MNOs in a service area and on the entry fee for which the DoT sent a letter to the TRAI for clarification. The TRAI responded in March 2009. However, even until December 2010, the DoT has neither accepted nor rejected the recommendations and no MVNO guidelines have been issued.

Mobile Value-added Service

Although the definition of Value-added Services (VAS) given in the UASL is generic and adequate, the TRAI defined VAS as 'Value-added services are enhanced services, in the nature of non-core services, which add value to the basic teleservices and bearer services, the core or basic services being standard voice calls, non-voice messages, fax transmission and bearer services'. (TRAI, January 2009).

In India, SMS, ringtone, and caller ring back tones (RBTs) constitute the bulk of the value-added services that the mobile operators are providing currently. The VAS delivery has so far been based on the SMS, Interactive Voice Response Systems (IVR), GPRS and wireless application protocol (WAP) portals' platforms. The VAS offerings are in areas such as entertainment, advertisement, gaming, contests such as interactive participation in television (TV) and radio games, reality shows, and news and support such as cricket alerts, news alerts, and travel alert details. With the introduction of 3G services, next-generation network (NGN)/converged network, this is expected to change in a big way as high-bandwidth multimedia content services, mobile TV and online gaming will push the demand for VAS as well as innovations in VAS products offerings.

The VAS market in India contributes to about 10–14 per cent of the total revenue of the mobile telecom service providers. There are multiple stakeholders in the VAS space. The commercial arrangements exist between telecom operators and VAS providers (VASPs) for providing these services. In many of these cases, the VASPs provide a technology platform which enables a user to access content on to his mobile or terminal device. In some of the cases the VASPs do not own the contents but they have arrangements with the content providers/content developers or copyright owners known as content owners. For some of the VAS, say SMS or messaging, the VAS platform including gateway/ middleware is provided by the telecom operator and VASPs only provide the content. In the commercial agreements, compliance to copyrights, digital rights management including sourcing of the content is the responsibility of VASPs. The various mobile VAS, be they voice-based or SMS-based, are provided to the mobile phone customers through the SIM card and through the short codes. The marketing of VAS is done through advertisement/media by telecom operators mainly for the contents hosted by them and also through the VASPs.

The VAS value chain consists of the following:

1. Telecom operators (access service providers)
2. Content aggregators/enablers (VASPs)
3. Content authors/producers or copyright owners (Content owners)
4. Device/mobile handset manufacturers
5. End-users, that is, customers

There is confusion and non-transparency in the structure and arrangements amongst the various stakeholders, which is cited as one of the main reasons for the limited growth of VAS in the country.

To clear the definition and to bring order and transparency in the industry, the TRAI released a consultation paper in 2008. The background and response to the consultation paper is highlighted by Sridhar and Raja (16 July 2008). For further readings on mobile VAS (MVAS), refer to Sridhar and Raja (4 January 2010).

After consultation with all stakeholders, the following were the highlights of the TRAI's recommendations for VAS (TRAI, January 2009):

1. The Authority was not in favour of a new licence category for VAS. However, the Authority recommended *registration* of VASPs or content aggregators, who wished to have common short code allotted by the DoT to provide VAS as 'Other Service Provider (OSP) – Value-added Services' under the OSP category. The registration could be for an LSA or on an all-India basis, covering all the LSAs.
2. Telecom access service provider shall provide fair access to telecom infrastructure to be independent of the content providers.

The regulator also recommended that the reconciliation on revenue-sharing between the network operators and VASPs should be left to them for their mutually agreeable terms and conditions.

Box 3.3 describes some of the innovative public MVAS that can be provided, especially in rural areas (Ratnoo and Sridhar, 29 April 2010).

Box 3.3   Public Utility of Mobile Value-added Service

With the Reserve Bank of India (RBI) increasing the daily transaction amount using mobile to Rs 50,000 there is a sudden spurt of activities in mobile banking with some of the major banks announcing a foray into this area. This is part of a trend that we are observing in MVAS which is moving away from being strictly entertainment-oriented towards being utility-based. Utility services can be defined as the ones that are of interest to a larger audience and which have transactional time value. With about half of our population subscribing to mobile services, what better way to disseminate public information than using MVAS.

The Indian MVAS industry valued at more than Rs 7,000 crores, has so far been dominated by entertainment services such as mobile music, RBTs, wallpapers and plain SMS. Though services such as RBTs have been a real success, the MVAS ecosystem players often paid little attention to other innovative services that can be of use to a larger audience. However, with hyper competition in the mobile services sector with voice ARPU continuing to decline, operators cannot afford to ignore public utility services. What are the examples of such utility services and how are they beneficial to the masses?

It would be very useful for the travellers if they are forewarned about the closing time of the Bandipur forest (in Karnataka) when they are planning to make a journey in the evening through the thickets, so that they do not get stranded at the entrance. Would it not be good if all of us are forewarned through SMS about the harmful effects of radiation emanating from the corona and the needed precautions to be taken during any solar eclipse which we may witness? Though one has been getting travel alerts from the traffic police in the metros whenever any major procession goes through the city, it would be of immense use to get alerts periodically; even better, based on the location in the city or town one is travelling in and the possible detours so that traffic bottlenecks can be circumvented. The SMS alerts targeted at specific locations may provide better reach than the traditional public address system to warn people of catastrophic events such as tsunami and flash floods that are quick and devastating. In countries such as Japan and Sri Lanka, technology known as 'cell broadcast' is used to send early alerts about earthquake and tsunami through mobiles to specific regions of the country. Similar early warning systems are being deployed in the coastal areas such as Kanyakumari district with the cooperation of the Indian Red Cross.

There are a host of health-related public VAS that can be provided, especially in the rural areas, to a select target audience. The SMS reminders and alerts can be sent to registered pregnant women for information on the next medical checkups; to mothers of infants on the next immunization schedule; to patients on the doctor's arrival time; and to nurses for attending and monitoring school health check-up programs in villages. For certain diseases such as tuberculosis, SMS alerts sent to patients on due dates for checkups, and to treatment providers on the next doses of medicine to be administered to specific patients, will improve the efficacy of directly-observed treatments.

The Tourism Promotion Councils in districts such as Kanyakumari in Tamil Nadu that earn substantial revenues from tourism can enable SMS-based advertisements for hotels in the area indicating room availability, prices and directions to promote a tourist-friendly hospitality industry.

Farmers from far-off villages often go to nearby towns and cities once a week to participate in *mandis* to sell their agricultural products. More often than not customers do not know the advantage of coming to the mandi and sometimes may not even know the venue and timings. Information on products and the mandi prices can be sent via SMS alerts so that customers have information and reason to come to the mandi to buy products. Increase in demand will also encourage farmers to come to the mandis to sell their products.

Mobile devices, which are truly personal, provide an opportunity to learn anytime anywhere. People in rural areas can be educated in languages using the 'word-a-day' paradigm through which they learn words, their meanings and pronunciations using voice SMS.

Mobiles may be the only channel available in certain rural areas of the country for enabling banking and can be deployed effectively using the strong distribution network of the mobile service providers. Mobile payment and banking enable quick turnaround time for remittances, especially in rural areas as demonstrated by service providers such as Zain in Africa.

Examples such as the ones mentioned above can be delivered by the ecosystem comprising mobile service providers, the appropriate government/ municipal agencies, and private content providers. So far, the various government agencies have not utilized the mobile services technology for disseminating public information. It is time that the agencies embrace this powerful and ubiquitous technology to disseminate public information for the benefit of the citizens. The service providers and content providers should also realize that utility MVAS remain an unexplored territory in India, however, show a good promise for the takeoff of the still fledgling MVAS industry in India.

Auctioning of Third-generation and Broadband Wireless Access Licenses

The TRAI released a consultation paper about 3G licenses way back in June 2006 (TRAI, June 2006). However, due to various political reasons, the 3G auctions were incessantly delayed and finally took place in April 2010. Box 3.4 (Prasad and Sridhar, 9 February 2009) highlights the delay in the procedure of allocating 3G licenses and the implications thereof.

Box 3.4   3G Delay Hurting the Economy

While 3G and BWA technologies usher in an era of mobile broadband at speeds up to 2 Mbps, the process of introducing these services by the government, unfortunately, is moving at the speed of a turtle! Incessant procedural delays, and drift between the finance ministry, telecom commission, the DoT and the TRAI on various issues including the reserve price have dashed the hopes of our country seeing the light of mobile broadband even by the end of the year. With elections around the corner, will it be wise to conclude that 3G is out for this year?

In the context of the present economic crisis, many countries are using telecom as a tool for improving economic activity. Adhering to its announced timeline, China issued 3G mobile service licenses to three state-owned operators on 7 January 2009. The operators are expected to spend about US$ 29 billion in building 3G telecom infrastructure. Chinese telecom equipment majors such as ZTE and Huawei expect their revenues to grow due to contracts with China Telecom, China Mobile and China Unicom. In the recession-struck US, President Obama wants to 'get broadband to every community in America', and the Congress is poised to include funding to upgrade and expand broadband Internet networks in the 'American Recovery and Reinvestment Plan'. Plans are afoot to give stimulus funds to increase high-speed Internet access across the country.

The broadband connections in India currently stand at 5.45 million, well below the target of 9 million (by 2007) as laid out in the broadband policy. Part of this dismal growth is due to the limitations for providing high-bandwidth connectivity using landlines, especially in the remote and rural areas of the country. Wireless broadband technologies provide an opportunity to expand broadband coverage in the country, to offer services such as telemedicine, e-government and e-education which in turn can spur socioeconomic development.

Further, the introduction of wireless broadband will increase the productive capacity of the Indian telecom sector, currently limited by the availability of the 2G spectrum and the excessive fragmentation of the industry. Our research shows that it is not just feasible, but also economically viable for the operators to use 3G infrastructure as a replacement/complement to their existing 2G infrastructure for voice services. In spectrum-scarce circles, the 3G spectrum will be used as an extension to 2G spectrum to fulfil the burgeoning demand for voice services at acceptable levels of quality.

Voice is a basic telecom service and the diffusion of voice services has been shown to have a positive impact on state-level Gross Domestic Product (GDP) in India. In the context of today's slowdown can we afford to dilly-dally on 3G while our growth declines?

It is ironical that the delay in rolling this essential service is being caused by the issue of the reserve price. The government seems to be taking the view that 3G provides high-end services, and therefore is unabashedly attaching great importance to the revenue-maximizing objective. But non-voice 3G services will be adopted mainly for development purposes in the initial years, provided the government gets its policy right.

The adoption of non-voice 3G services by the mainstream market, on the other hand, will be far from immediate. Even though operators are offering non-voice services on the 2G networks, India remains a voice-centric market. The ARPU in Indian metros and Category A circles is about Rs 274/month. Out of this, data services contribute roughly 12 per cent. The SMS contributes about 50 per cent of the non-voice revenue of the operators. Even in an advanced market like the US, non-voice contributes only about 16 per cent to the overall ARPU of the operators. The only exceptions are the Japanese and South Korean markets which have data ARPU contributing to more than 40 per cent of revenues. However, it is well known that Japanese and Korean markets lead the rest of the world markets of the world in terms of advanced mobile services by two to three years.

Therefore 3G should be treated as an input for the basic service of voice. If the price of the 2G spectrum was kept low on account of it being a basic service, then the 3G spectrum should also be priced low. On the contrary, the nil reserve price recommended by the TRAI in 2005 was modified later to Rs 1,100 crores in 2006 (for pan-India); and was increased to Rs 2,020 crores by the DoT in 2008. After the Finance Ministry increased it to Rs 4,040 crores, the DoT agreed to fix it at Rs 3,540 crores. What is the rationale behind these arbitrary increases? Our research using data on the international pricing of the 3G spectrum indicates that the reserve price should be in the range of Rs 850 crores for pan-India allocation.

The unviability of 3G also stems from the fragmentation of the 3G spectrum which is being auctioned in 5-MHz blocks as opposed to larger blocks that would enable economies of scale to be reaped. The international average for 3G spectrum allocation is 15–20 MHz per operator.

Thus the government is delaying the 3G auction to set a high price, which becomes daily more unviable as the salubrious effects of the introduction of a new technology are lost. In short, time lost is a drain on growth, a blow to economic development, and a hurdle for the diffusion of basic voice service in the country.

Finally, the Indian government announced the policy for 3G mobile services in August 2008 (DoT, August 2008a). In line with the TRAI's recommendation (TRAI, 2006), the government opted for a simultaneous ascending auction for the allotment of a start-up spectrum of 2×5 MHz in the 2.1 GHz band with specified reserve prices for different categories of LSAs. Note that 2×5 MHz is the minimum carrier requirement for providing 3G services using WCDMA technology in the 2.1 GHz band. With broadband penetration at a very low level, the Government of India took a proactive step in introducing the BWA service in the country. The policy for auctioning spectrum for BWA was scheduled to start two days after the 3G spectrum auction (DoT, August 2008b). The auction procedure for 3G and BWA spectrum is described in Figure 3.11 and Table 3.8.

Each auction was a simultaneous ascending e-auction, run over the Internet. More specifically, each of the auctions involved the following two-stage process:

1. *Clock Stage*: A clock stage would establish the bidders to be awarded a block in each telecom service area where there is at least one block available to auction. In this stage, in each service area, bidders would bid for a block (that is, a right to a single spectrum block not linked to any specific frequency). The first clock round will begin with all lots in a circle being offered at the specified reserve price. The clock stage would consist of a number of rounds (the 'clock rounds'). These rounds would stop once (i) for every service area where spectrum is available, the number of bids at the prices set in the last completed Clock Round is less than or equal to the number of blocks available; and (ii) there are no opportunities for bidders to increase their demand as allowed by the activity rules. The activity rules are determined by the eligibility points the bidders receive. Each service area is given a point-based weightage. A bidder cannot bid for more service areas than its eligibility points allow. The bidder's earnest money deposit determines its eligibility to bid in the first round. In subsequent rounds, eligibility points are determined by the bidding activity in the previous round.

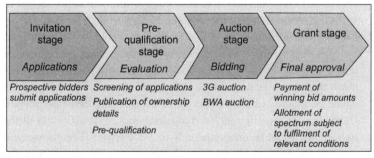

Figure 3.11   The 3G/Broadband Wireless Access Auction Process

Table 3.8   The 3G/Broadband Wireless Access Auction Process Parameters

Parameter	Spectrum in 2.1 GHz (for 3G)	Spectrum in 800 MHz (for 3G)	Spectrum in 2.3 GHz (for BWA)
Who is eligible?	Incumbent UAS or cellular mobile telecommunication services (CMTS) licensees; New entrant needs to acquire either of these licenses	Incumbent UAS-CDMA operators	Incumbent UAS or CMTS licensees; Category A or B type ISPs; new entrant needs to acquire one of the above licenses
Block auctioned	Three/four blocks of 2×5 MHz depending on availability of spectrum in each LSA; one block given to government operators BSNL and MTNL	One block of 2×1.25 MHz depending on availability in the LSA	Two blocks of 20 MHz (unpaired) depending on availability of spectrum in each LSA; one block given to government operators BSNL and MTNL
Reserve Price for auction (in million of INR)	Delhi and Mumbai: 3200; Kolkata: 1,200; Category A: 3,200; Category B: 1,200; Category C: 300	Delhi and Mumbai: 800; Kolkata: 300; Category A: 800; Category B: 300; Category C: 75	Delhi and Mumbai: 1,600; Kolkata: 600; Category A: 1,600; Category B: 600; Category C: 150

Source: DoT, October 2009.

The bidding activity is determined by new bids, plus any provisional winning bids that are held through the clock round. Bidders do not initially have to be active on all their target services areas, but the activity requirement tightens in stages (80 per cent, 90 per cent, 100 per cent). This would establish a common winning price for all blocks within a service area, and the winning bidders in each service area;

2. *Frequency Identification Stage*: The clock stage would be followed by an assignment stage that would allocate specific frequencies available to the winning bidders identified in the clock stage in case of service areas where more than one spectrum block is being auctioned. The assignments would be announced simultaneous with the outcome of the clock stage. In the case of the 3G and BWA auctions, the initial allocation of the frequencies would be done randomly by the software. The government reserves the right to change this allocation over the duration of the licence period where this may promote spectrum efficiency.

Winning bidders would pay the sum of the relevant winning prices set in the clock stage for circles in which they are allocated a block. All winning bidders in a service area would have equal winning clock stage bids as a consequence of the Auction Rules (for details on the auction procedure, refer to the DoT (August 2008).

The 3G auction was conducted over thirty-four days (from 9 April 2010 to 19 May 2010) and involved 183 rounds of bidding across the LSAs. All the seventy-one blocks of spectrum available were put up for auction in the twenty-two LSAs in the country, leaving no unsold lots. The BWA auction was conducted over sixteen days from 11 June 2010 and was completed in 117 rounds.

Table 3.9 gives the details of the winners and the winning bid prices for each of the twenty-two LSAs:

The following observations can be made from Table 3.9:

1. The table indicates that the amount accrued to the government due to the 3G and BWA auction was Rs 106,223.53 crores including Rs 16,750.78 crores for 3G and Rs 12,847.27 crores for BWA respectively for pan-India coverage by the government operators BSNL and MTNL.
2. The final pan-India 3G and BWA bid amounts were five and seven times the corresponding reserve prices. It is to be noted that the BWA reserve price was half the 3G reserve price. On an average, the BWA bid amount in each circle was about 68 per cent of the corresponding 3G bid price.
3. The final pan-India BWA bid amount was about 76 per cent of the final 3G bid amount.
4. For the first time, the licence was separated from the spectrum allocation procedure.

Comparison of Mobile Services in China

China is the world's largest mobile communications market. The mobile subscriber base at the end of 2010 was about 860 million (Marbridge, 2011). The fixed-line subscriber base continues to decline as has been happening in India and is shown in Figure 3.12. Government operators provide mobile telecom services in China in thirty-one regions of the country.

Despite this substantial growth of mobile services the penetration of mobile services is not uniform across different areas of the country.

Table 3.9   Results of the 3G/Broadband Wireless Access Spectrum Auction

S. No.	Licensed Service Area	3G			Broadband Wireless Access		
		Bid Amount (in Rs crore)	No. of Win- ners	Total Bid Amount (in Rs crore)	Bid Amount (in Rs crore)	No. of Win- ners	Total Bid Amount (in Rs crore)
1	Delhi	3,316.93	3	9,950.79	2,241.02	2	4,482.04
2	Mumbai	3,247.07	3	9,741.21	2,292.95	2	4,585.90
3	Kolkata	544.26	3	1,632.78	523.20	2	1,046.40
4	Maharashtra	1,257.82	3	3,773.46	915.64	2	1,831.28
5	Gujarat	1,076.06	3	3,228.18	613.85	2	1,227.70
6	Andhra Pradesh	1,373.14	3	4,119.42	1,059.12	2	2,118.24
7	Karnataka	1,579.91	3	4,739.73	1,543.25	2	3,086.50
8	Tamil Nadu	1,464.94	3	4,394.82	2,069.45	2	4,138.90
9	Kerala	312.48	3	937.44	258.67	2	517.34
10	Punjab	322.01	4	1,288.04	332.27	2	664.54
11	Haryana	222.58	3	667.74	119.90	2	239.80
12	Uttar Pradesh (W)	514.04	3	1,542.12	183.37	2	366.74
13	Uttar Pradesh (E)	364.57	3	1,093.71	142.50	2	285.00
14	Rajasthan	321.03	3	963.09	97.32	2	194.64
15	Madhya Pradesh	258.36	3	775.08	124.66	2	249.32
16	West Bengal	123.63	4	494.52	70.97	2	141.94
17	Himachal Pradesh	37.23	3	111.69	20.66	2	41.32
18	Bihar	203.46	4	813.84	99.28	2	198.56
19	Orissa	96.98	3	290.94	63.63	2	127.26
20	Assam	41.48	3	124.44	33.02	2	66.04
21	North East	42.30	3	126.90	21.27	2	42.54
22	Jammu and Kashmir	30.30	4	121.20	21.27	2	42.54
	Total	16,750.58		50,931.14	12,847.27		25,694.54
	BSNL/ MTNL To Pay	16,750.58		16,750.58	12,847.27		12,847.27
	Grand Total			67,681.72			38,541.81

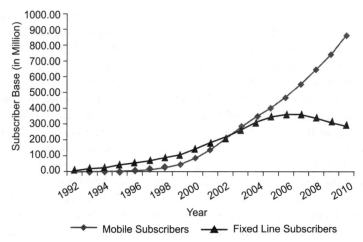

Figure 3.12   Growth of the Telecom Subscriber Base in China

Though India was late in introducing mobile services, the annual growth rate has been consistently higher than China from the year 2000 onwards. India and China account for more than 25 per cent of the total mobile subscriber base in the world.

The Ministry of Post and Telecommunications (MPT) of the Chinese government was the monopoly mobile service provider in the country. In July 1994, the operational arm of MPT was renamed the China Telecommunications Corporation (China Telecom). China Unicom Ltd., a subsidiary of China United Telecommunications Corporation, was set up at the same time by the former Ministry of Electronics and Information to foster competition in the telecom industry. In May 2000, China Telecom was split into four separate entities, one of which was the lucrative mobile services, christened as China Mobile Communications Company (Loo, 2004). Until recently China Mobile and China Unicom were the two licensed mobile operators in China in the thirty-one provinces. China Mobile is the world's largest mobile phone company in terms of number of subscribers and has about 60 per cent of mobile subscriber base in the country. China Unicom is the second largest mobile telecom network operator. China Unicom started its GSM services in 1995, CDMA service in 2002 and accounts for about 33 per cent of mobile subscriber base in the country. (PBC, 2006).

After the restructuring of China Telecom in 2000, the fixed-line services of China Telecom were split into two competing companies operating in the north and south of the country. The company operating in the south was christened China Telecom. The other company, China Network Communications Group Corporation (China Netcom) covers ten provinces in the north including the capital Beijing. Though China Telecom and China Netcom were licensed to provide basic fixed-line services, they started participating in the mobile services market through Xiaolingtong (also referred to as 'a Little Smart', abbreviated as XLT) based on the personal handy-phone system (PHS) technology. From its 1998 launch in Yuhang, a small city in the Zhejiang province, XLT has spread to more than 600 cities nationwide with its subscriber base reaching about 80 million in 2005 (Yuan *et al.*, 2006). By connecting to existing fixed-line telephone networks through micro-cell radio, XLT provides mobile wireless access to ordinary local PSTN services. Though the government officially restricted XLT to outlying cities and provinces, the service has gained a strong national market foothold (PBC, 2006). The XLT is registered as a fixed-line service and by 2003, accounted for more than 40 per cent of all fixed-line subscriber additions in China (PBC, 2006). In 2009, China Telecom was allowed the provisioning of mobile services.

In July 2009, the Ministry of Industry and Information Technology of China awarded 3G licenses. China's biggest mobile operator, China Mobile, was awarded a licence for TD-SCDMA (Time Division–Synchronous Code Division Multiple Access), the domestically developed 3G standard. The other two main carriers, China Telecom (erstwhile provider for fixed-line telecom services) and China Unicom, received licenses for CDMA2000 and WCDMA, respectively. Just before the licenses were awarded, China Unicom merged with China Netcom, thus leading a consolidated three-player market consisting of only China Mobile, China Telecom and China Unicom. China's 3G users numbered 47 million at the end of 2010; China Mobile continues to lead the pack with 20.7 million 3G users, followed by China Unicom at 14.06 million and China Telecom at 12.29 million users. China Mobile is also experimenting with TD-LTE as a possible long-term migration path from TD-SCDMA in partnership with Qualcomm, USA and to lead the 4G evolution in China.

Until 2001, the average monthly cellular subscription charges in China were 50 Yuan (US$ 6.00) and a three-minute local call was priced at 1.20 Yuan (US$ 0.15) (ITU, 2003). Though the competition is limited in China, the battle for market share has caused aggressive price cutting by China Mobile and China Unicom. The two mobile operators started offering promotional prices in several provinces that were well below those permitted under the prescribed tariff levels. In October 2001, the Ministry of Information and Industry (MII) in China ordered the two companies to halt their price war and threatened to punish company officials who offered rates below state-sanctioned levels (PBC, 2006). According to the MII, the actual rates paid by China's mobile phone subscribers fell by 20 per cent on an average in 2002 due to the use of various discount plans. Price wars continued into 2003, when China Unicom reduced its rates for incoming calls in some areas. After the launch of XLT the GSM and CDMA services offered by China Mobile and China Unicom were under severe threat. They had to reduce prices further to retain customers. The price war slowed in 2004 after the authorities imposed price regulations that required operators to gain approval for any new tariff promotional plan (PBC, 2006). The rules required that mobile carriers and XLT providers must charge higher rates than fixed-line carriers. China still has the 'receiving party pays' regime where the mobile operators charge their own subscribers for termination of received calls.

All the operators in China are government-owned. There is a limited competition in mobile services. However, the introduction of XLT services by China Telecom and China Netcom with the low calling charges and free incoming calls is having an impact on the traditional GSM and CDMA services offered by China Mobile and China Unicom. Regulatory levies in China and India are shown in Table 3.10. The levies in China are much lower compared to India, partially due to the presence of only government operators.

These high licence fees erode the profit margin of the Indian operators. Hence Chinese companies have much higher returns on capital employed,

Table 3.10   Comparison of Regulatory Levies in India and China

Parameter	China (as percentage of revenue)	India (as percentage of gross revenue)
Service Tax	3 per cent	More than 12 per cent
Licence Fee	Nil	6–10 per cent (including 5 per cent U)
Spectrum Charges	About 0.5 per cent	2–6 per cent
Total Charges	About 0.5–3 per cent	About 17–26 per cent

Source: TRAI, 2005b.

22.87 per cent, compared to Indian mobile operators who have returns of only about 7.83 per cent (TRAI, 2005c). Based on the discussions above, the characteristics of the Chinese and Indian telecom services market are summarized in Table 3.11.

Table 3.11   Summary of the Chinese and Indian Mobile Telecom Market

Characteristic	China	India
Market structure	Duopoly with Government Operators until 2009; currently three operators	Highly competitive (up to ten to fourteen operators) with just two government operators
Price level	About the same	
ARPU	ARPU in China is marginally better than in India	
Minutes of usage (MoU)	MoU in India is marginally higher than in China	
Return on capital employed (RoCE)	RoCE of companies in China is better than India	
Regulatory Levies	Low	High

4

Spectrum Management for Mobile Services in India
A Conundrum

Spectrum is an essential resource for the provision of mobile services. The management of radio frequency spectrum is important to meet the following objectives: (i) granting of exclusive rights to spectrum (ii) ensuring its efficient use, and (iii) promoting competition in services (Falch and Tadayoni, 2004). In addition, in emerging economies spectrum management must achieve the development goals of universal telecom access at low prices. As pointed out by Falch and Tadayoni (2004), the focal point of regulation of scarce resources, namely spectrum, has changed from pure coordination and planning to the creation of a competitive and sustainable environment for various telecom services. In this chapter, we critically look at the history of spectrum management in India with respect to allocation and pricing.

Stages in the Management of Spectrum

Before the introduction of mobile services, spectrum intended for commercial usage in the 800 MHz, 1800 MHz and 1900 MHz frequency bands was entirely in the control of the national defence force of India. The utilization of spectrum for commercial purposes began with the release of a limited amount of spectrum in 1995. The management of spectrum in the country can be divided into four stages.

First Stage: Auctioning Scarce Spectrum (1995–2003)

The Indian government auctioned 2×4.4 MHz (paired frequency division duplex spectrum assignment) for GSM technology in the 890–915 MHz band paired with 935–960 MHz in 1995. Two operators were selected for each LSA (for details on LSAs in India, refer to Prasad and Sridhar, 2008). Subsequently, the third operator licence was awarded along with 2×4.4 MHz of start-up spectrum in the same 900 MHz band to the government operator on a pro bono basis in 2001. The fourth operator licence was issued in 2001 using a three-stage auction procedure and a start-up spectrum of 2×4.4 MHz in 1710–1785 MHz paired with 1805–1880 MHz band was given to the winning bidder. In addition to the entry fees, licensees were required to pay a percentage of annual revenue as spectrum charges.

Further allocation of spectrum beyond the start-up spectrum levels was based on availability and justification and attracted additional revenue share as spectrum charges. The contractual rights of spectrum holders were incrementally established through a series of government orders. In 2002, the subscriber-linked spectrum allotment procedure, referred to as Subscriber-based Criterion (SBC) was introduced which laid down a roadmap for allotment up to 2×12.5 MHz of spectrum per operator in each LSA (see Table 4.1).

This method of spectrum allocation was very different from the methods followed in other countries where a sizable spectrum block (about 2×15 MHz) was given to the operators as start-up spectrum (TRAI, May 2005). The rationale for adopting a different approach was the scarcity of spectrum due to non-availability from the Department of Defence.

Table 4.1 Spectrum Allocation Criteria in 2002

Quantum of Spectrum Allotted	*Minimum Subscriber Base Required (million)*	*Annual Spectrum Charges (as per cent of Adjusted Gross Revenue)*
2 × 4.4 MHz	–	2
2 × 6.2 MHz	–	3
2 × 8 MHz	0.5	3
2 × 10 MHz	1.0	4
2 × 12.5 MHz	1.2	5

Source: Sridhar, 2007.

Second Stage: De-linking Licence from Spectrum (2003-6)

After a couple of years of litigations between the BTS operators and the GSM mobile operators as discussed in Chapter 2, the Indian government announced the UASLs in November 2003 that allowed basic service licence holders to provide full mobility-based services with a stipulated entry fee based on the bid price paid by the fourth operator in 2001 (DoT, 2003). The fixed fee-based licence (as opposed to auction-based) theoretically allowed any number of mobile licenses to be provided and implicitly de-linked spectrum allocation from licensing. Though firms were awarded licence after paying the required entry fee, they were given start-up spectrum only as and when available. Following the entry of two to three CDMA-based mobile operators in each LSA, one to two new firms also paid the stipulated entry fee and got licenses to operate GSM services in certain LSAs.

In 2005, the TRAI reviewed the spectrum allocation process taking into account spectrum availability and efficient techniques for utilization of assigned spectrum (TRAI, 2005). A maximum of 2×10 MHz is being allotted to GSM operators while the world average is about 2×20 MHz. Taking into account the frequency spectrum used by the Ministry of Defence for its navigational aids and by the Railways, additional spectrum allocation in the 900 MHz was not possible. Hence any additional spectrum for the first three operators also has to be allotted in the 1800 MHz frequency band only. However, the 1800 MHz frequency band is extensively used by the Air Force and Army networks and hence had not been released fully to GSM operators. A maximum of 2×5 MHz is allotted in the 800 MHz band for the CDMA operators while the world average for CDMA operation is about 2×15 MHz. The Defence uses the 1900 MHz band and hence it is not currently available to either GSM or CDMA operators. The Defence has also been coordinating the band (1880–1900 MHz paired with 1970–1990 MHz) for indigenously developed CorDECT fixed wireless local loop technology.

Analysing the above, the TRAI (2005) stated that the spectrum held by the GSM and CDMA operators was well below international averages. It was recommended that existing operators be given adequate spectrum before considering allocating spectrum to new service providers especially since 'there is adequate competition in almost all service areas'. The TRAI (2006) continued to maintain that there was a shortage of 2G spectrum.

The entitlement of incumbents naturally extended to the future 3G spectrum since in a spectrum-scarce environment 3G could be considered as an extension of 2G.[1] It also followed that the scarce resource of spectrum need not be spread too thin thereby justifying a cap on the number of operators.

Third Stage: Stricter Criteria for Allocation of Spectrum (2006–8)

As technological progress took place, it began to be believed that if used maximally, spectrum held by incumbents was sufficient for meeting their near-term requirements. Meanwhile additional spectrum was also getting vacated by the Defence. As a result, in keeping with the principle of maximal usage of spectrum, new subscriber-based norms were defined, incumbents kept out of fresh allocations, 3G treated as a separate service from 2G, and the cap on the number of operators removed. As mentioned in the TRAI recommendation of 2006 on 3G spectrum (TRAI, 2006), 'the Defence services have agreed to vacate 2×20 MHz in the 1800 MHz band, in addition to 25 MHz in the 2.1 GHz UMTS band. The availability of additional spectrum in the 1800 MHz band provides sufficient room for growth of 2G services for the medium term.' Therefore 'the Authority has recommended that the Government should not treat the allocation of 3G spectrum in continuation of 2G spectrum'.

The TRAI (2007) recommendation that no cap be placed on the number of telecom access providers in the country allowed more new firms to enter the market by paying the low fixed entry fee. It even allowed CDMA operators to obtain GSM licenses. The new licensees were put in queue for assignment of spectrum as and when available. The TRAI (2007) further stated that the spectrum allocation criteria should take into account methodologies such as synthesized frequency hopping, and use of advanced codecs for efficiently utilizing allotted spectrum. Following this, the government tightened the SBC for additional spectrum allocation to existing operators making them ineligible for participating in the new allocation of spectrum. Table 4.2 gives the SBC released by the government in 2008, which are much stricter than those shown in Table 4.1.

Table 4.3 illustrates the spectrum allocation in different LSAs.

---

[1] The 3G spectrum is designed for providing connectivity at speeds of up to 2 Mbps and hence is capable of providing services such as mobile TV, and real-time video conferencing while 2G spectrum is used for low-bandwidth services such as voice and SMS.

Table 4.2 Subscriber Base Criteria for Additional Spectrum Allocation: 2008–Current

Quantum of Spectrum Allotted	Minimum Subscriber Base Required (million)	Annual Spectrum Charges (as per cent of Adjusted Gross Revenue)
For GSM Services		
2 × 4.4 MHz	–	2
2 × 6.2 MHz	0.5–0.8*	3
2 × 7.2 MHz	1.5–3.0	
2 × 8.2 MHz	1.8–4.1	3
2 × 9.2 MHz	2.1–5.3	
2 × 10.2 MHz	2.6–6.8	4
2 × 11.2 MHz	3.2–6.8	
2 × 12.2 MHz	4.0–9.0	5
2 × 14.2 MHz	5.7–10.7	5
2 × 15 MHz	6.5–11.6	6
For CDMA Services		
2 × 3.75 MHz	0.15–0.40	2
2 × 5.0 MHz	0.5–1.2	2

Source: WPC, 2008.
Note: * Range across the LSAs.

Fourth Stage: Introduction of Third-generation and Broadband Wireless Access (2010–)

The government announced the policy for 3G mobile services in August 2008 (DoT, August 2008). In line with the TRAI's recommendation (TRAI, 2006), the government opted for a simultaneous ascending auction for allotment of a start-up spectrum of 2×5 MHz in the 2.1 GHz band with specified reserve prices for different categories of LSAs. Details of the auction procedure are explained in the previous chapter. Note that 2×5 MHz is the minimum carrier requirement for providing 3G services using WCDMA technology in the 2.1 GHz band. The 3G policy also states that 2×1.25 MHz carriers will be allotted to UAS-licensed CDMA operators at a price equal to the highest bid received for the 2.1 GHz band, prorated for 2×1.25 MHz. The adage is that market-based assignment mechanisms such as auctions are superior from an economic efficiency point of view in allocating scarce resources such as electromagnetic spectrum. However, Gruber (2007) points out that in Europe, the number of firms that introduced 3G services was less than the number of firms that received the

Table 4.3 Spectrum Allocation in Different Licence Service Areas in India (as of January 2009)*

S. No.	LSA	GSM				CDMA		
		Total Quantity of Spectrum Allotted (MHz)	Number of Operators Allotted Spectrum (Average in MHz Allotted†)	Number of Licenses Waiting for Spectrum	Future Spectrum Availability (MHz)	Total Quantity of Spectrum Allotted (MHz)	Number of Operators Allotted Spectrum (Average in MHz Allotted)	Future Spectrum Availability (MHz)
Category: Metros								
1	Delhi	53.6	7 (7.8)	5	–	15	4 (3.75)	2.5
2	Mumbai	72.6	11 (6.6)	–	–	15	4 (3.75)	2.5
3	Kolkata	60.4	10 (6.2)	–	15	13.75	4 (3.75)	2.5
4	Chennai	66.6	11 (6.2)	–	–	13.75	4 (3.75)	3.75
Category A								
5	Maharashtra	63	11 (6.0)	1	–	15	4 (3.75)	1.25
6	Gujarat	51.6	9 (5.8)	2	–	12.5	4 (3.75)	5
7	Andhra Pradesh	68.6	12 (5.8)	–	15	13.75	4 (3.75)	2.5
8	Karnataka	69.4	12 (5.8)	–	15	15	4 (3.75)	2.5
9	Tamil Nadu	67	11 (6.2)	–	15	12.5	4 (3.75)	5
Category B								
10	Kerala	61.2	11 (5.6)	–	15	15	4 (3.75)	2.5
11	Punjab	50	9 (5.6)	3	–	15	4 (3.75)	1.25

12	Haryana	59.4	11 (5.4)	1	–	12.5	4 (3.75)	5
13	Uttar Pradesh (West)	61.2	11 (5.6)	–	–	13.75	4 (3.75)	3.75
14	Uttar Pradesh (East)	52.4	9 (5.4)	2	–	13.75	4 (3.75)	3.75
15	Rajasthan	63.8	12 (5.4)	–	–	15	4 (3.75)	–
16	Madhya Pradesh.	63	11 (5.4)	–	15	12.5	4 (3.75)	3.75
17	West Bengal	53.2	10 (5.4)	–	–	11.25	3 (3.75)	5
Category C								
18	Himachal Pradesh	57.6	11 (5.4)	–	–	10	4 (2.5)	6.25
19	Bihar	66.8	12 (5.8)	–	–	13.75	4 (3.75)	3.75
20	Orissa	59.4	11 (5.4)	–	15	11.25	3 (3.75)	6.25
21	Assam	55	10 (5.6)	1	–	10	4 (2.5)	7.5
22	North East	53.2	10 (5.4)	1	–	10	4 (2.5)	7.5
23	Jammu and Kashmir	49.4	10 (5.0)	1	–	12.5	5 (2.5)	5

* The spectrum allocation is one-way (uplink or downlink) of the paired spectrum.
† Rounded off to the next higher GSM carrier level (multiples of 0.2 MHz).

licenses. It has to be seen whether the 'winner's curse' as experienced in the 1995 licensing in India and later in Europe in 2001 will be repeated.

As can be seen, the BWA auction amount was about 70 per cent of the 3G auction amount, though the reserve price of the BWA auction was just 50 per cent of the 3G auction reserve price. However, it is to be noted that the BWA start-up spectrum was almost twice that available for 3G auction.

Critique of Spectrum Policy

The following issues need to be addressed in the policies for allocation and pricing of spectrum, details of which are also explained in Prasad and Sridhar (20 September 2008) and Sridhar and Prasad (2011).

The Legacy of Subscriber-based Criteria

The SBC is contentious and intractable, especially with the growth of data services. Further, as new spectrum bands are being made available for various wireless telecom services such as mobile TV, and BWA, operators are likely to be holding several bands simultaneously, and servicing subscribers using a combination of bands. It will not be possible to segregate subscribers into different bands in order to determine the subscriber numbers for a particular band for applying the SBC. Fortunately, the need to have SBC in place to assign spectrum has dwindled. In 2002, the year in which the SBC were announced, there were four to six operators in every circle. In 2010, the average number of licensees in each circle has gone up to fifteen. Any inefficiency in the use of spectrum is sure to be penalized by market forces and does not need to be administratively monitored. Hence the SBC methodology for additional spectrum allocation should be dispensed with and a market-oriented approach to spectrum allocation such as auction should be instituted.

Annual Spectrum Usage Charge

As indicated, the annual spectrum charges vary depending on the amount of spectrum held. The escalating percentage is a way of charging for additional spectrum which currently does not come with an upfront payment. It is also meant to bring about the efficient use of spectrum. With a market-oriented mechanism such as an auction for allocating additional spectrum, escalating annual spectrum charges are no longer required. However, the government may also desire that the migration to uniform rates of usage charges should be revenue-neutral. In view of the additional revenues accruing from auction-based allocation as well as transfer charges, the government need not worry unduly about the loss of revenue on account of the annual usage charge. Sridhar (2006) argues that regulatory levies such as spectrum charges should be equivalent to the cost of administration.

However, if there are political compulsions to maintain revenues at previous levels then one option would be to take the average usage charge rate for the industry and fix this as a uniform rate.[2] Uniformity of annual spectrum charges across all frequency bands and technologies would facilitate change of use and effective use of allocated spectrum.

Trunking Efficiency and Economies of Scale

It is well known in telecom that the trunking efficiency, which is the ratio of the average utilization of a network to its peak capacity for a given grade of service, goes up as the peak capacity increases. In the context of wireless systems, this means that the capacity of a network increases as the number of the associated spectrum blocks decreases keeping the total spectrum available the same. It is important to note that the incremental increase in spectral efficiency levels off beyond a certain size of spectrum allotment. The absolute value of this saturation point depends on the technology and service (messaging/telephony/data) provided. However, from a regulatory standpoint, it is important to create a market situation wherein most operators have sufficient spectrum to be operating at or above the saturation point. Figure 4.1 illustrates the trunking efficiency principle as described above.

As shown in Figure 4.1, the effective network capacity decreases as the total amount of 40 MHz spectrum is split amongst more and more operators.

In the parlance of economics, the presence of trunking efficiency implies that the telecom industry exhibits 'economies of scale'. Prasad and Sridhar (2008) estimate the degree of scale economies in the Indian mobile industry and find significant scale economies even in Category C LSAs that comprise a large rural sector. When economies of scale are present the presence of a large number of firms beyond a certain threshold results in increasing per unit costs. Thus social welfare is maximized when

[2] So far as existing GSM operators are concerned, each one of them holds a different quantum of spectrum and pays spectrum charges at different rates of AGR in different service areas. The CDMA players pay a uniform 2 per cent as indicated in Table 4.2.

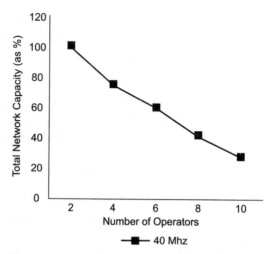

Figure 4.1   Illustration of Trunking Efficiency

there are a few firms producing at efficient scales of operation and whose competition with each other restricts mark-ups over costs.

Optimal versus Maximal Use of Spectrum

The formulation of the spectrum policy in India began under conditions of very limited availability of spectrum. In the initial phase when spectrum was scarce the policymaker's requirement of maximal usage of spectrum with the associated subscriber-based norms was justified. However, as spectral efficiency increased and additional spectrum got released, more nuanced definitions of efficiency needed to be applied in order to promote the growth of the industry.

The single-minded agenda on maximizing the number of subscribers per unit of spectrum ignores the importance of efficiently using other inputs like BTSs. The specific interpretation of technical efficiency used by the government has resulted in high reuse of spectrum and hence more BTSs and cell towers. In some circles, the inter-site distance between cell towers is less than 100 metres, which is one of the lowest in the world leading to iron-clad structures clogging some of the cities. The increase in BTSs increases the capital and operational cost of the service providers thus leading to possible inefficiencies.

The production function estimated in Prasad and Sridhar (2008) clearly indicates the presence of economies of scale in the industry. This implies that unit costs decline with an increase in the scale of operation, at least over the range of subscriber minutes observed in the industry to date. In view of this, the policymaker should allow accumulation of spectrum where such accumulation leads to lower costs without increased threat of cartelization. This would allow operators to integrate operations and aggregate spectrum holdings to take advantage of economies of scale. An article on the concept of efficiency in mobile services is given in Box 4.1 (Prasad and Sridhar, 1 June 2008).

Box 4.1   Spectrum Policy Too Narrowly Focused?

Economics, Lionel Robbins famously stated, is the study of the allocation of scarce resources among alternative ends. Policymakers could be forgiven for thinking that spectrum was the only scarce resource in the industry and that mobile phone operators were the only users of spectrum. This is a consequence of a policy debate narrowly focused on optimizing the allocation of spectrum within the mobile industry, without regard to patterns of use of other inputs and of the usage of spectrum by much larger users of spectrum outside the mobile industry.

In mobile communication, the amount of spectrum possessed by an operator and the number of BTSs deployed at the cell sites, directly impact the subscriber base and the traffic the operator can support. The operator maximizes profit by using a judicious combination of the two main inputs, namely, spectrum and BTS, taking into account their relative prices. This is often referred to as 'allocative efficiency' in economic literature. The other equally important concept of efficiency is 'technical efficiency' that requires the firm to produce the maximum output (that is, support maximum number of subscribers) given the level of inputs. In order to achieve economic efficiency, a combination of allocative and technical efficiencies should be achieved.

The history of the spectrum policy in India as implemented by the DoT has defined technical efficiency in a very narrow way and ignored allocative efficiency altogether. The approach to frequency management has mainly focused on extracting the maximum use from a given spectrum while ensuring a minimum quality of service. Eligibility for incremental blocks of spectrum is dependent on fulfilling subscriber-based norms that specify threshold limits in line with the maximum number of subscribers that can possibly be serviced by spectrum already held. But maximal usage of spectrum is only one possible technically efficient input combination. There are several others with lower subscribers per unit of spectrum but higher subscribers per BTS.

The specific interpretation of technical efficiency used by the DoT has resulted in high reuse of spectrum and hence more BTSs. In some circles, the inter-site distance between cell towers is less than 100 metres, which is one of the lowest in the world leading to these structures clogging our cities. Apart from cluttering the landscapes the antennas mounted on the towers radiate radio frequencies that may be harmful to city dwellers. The BTSs consume electricity to the tune of about 1 KWH for a minimal configuration, requiring the operators in cities such as Delhi to spend as much as Rs 10 crores per annum on electricity charges.

The notion of technical efficiency used by the policymaker would conform to allocative efficiency if the abundance of spectrum relative to the other inputs mandated that it be used maximally, or if the productivity of BTSs was sufficiently high relative to the spectrum. This condition need not hold. Further, even if it may be relevant at the early stages of an industry it may not continue to be relevant if conditions change. For example, technological advances that increase the productivity of spectrum would mandate using spectrum in higher proportions. Today while in India the mean holding of spectrum is only about 2×7 MHz, it is 2×21.5 MHz across 140 GSM operators in forty countries.

This trajectory of spectrum management has resulted in operators becoming allocatively inefficient, as they are forced to use more of a relatively expensive input (in this case BTSs). Thanks to the above phenomenon the cell tower business has prospered in India leading to astronomical valuations. However, operators are being deprived of spectrum and rendered incapable of reaping the advantages of economies of scale for themselves and for the industry.

An important lacuna in spectrum policy stems from the fact that there is no integrated management of spectrum. A bulk of the spectrum is controlled by the armed services who pay nothing for their utilization. This leads to hoarding by the armed forces and a scarcity of spectrum for commercial uses.

That all users must use inputs in line with their relative scarcity as represented by shadow prices is not merely a narrow commercial consideration motivated by profit maximization but a broader injunction for efficient allocation of social resources using the organizing power of the invisible hand of markets. Critics of the market mechanism rightly point out that the market typically does not pay heed to considerations of equity. However, as it happens, the saturation of mainstream markets and emergence of purchasing power in previously excluded pockets is forcing operators to search for the gold at the lower reaches of the economic pyramid. While market forces seem to be adequately addressing the social objective of equity, they are being hobbled in generating efficiency by wrongheaded policy formulation.

Efficient Use of Mobile Infrastructure

Under-utilization of BTSs leads to an increased demand for spectrum. There are several new technologies to ensure optimal yield from BTSs. These include:

1. Implementation of in-building solutions such as Femtocell to improve the efficiency of BTSs.
2. Single-antenna Interference Cancellation that can improve the downlink bandwidth of GSM networks without changes to network configurations.
3. Utilization of smart antenna arrays that can confine channels into a narrow beam thus improving capacity gain.

The government should monitor the optimal use of base stations in the same way as it monitors the use of spectrum. Prescribing subscriber-based norms for BTSs is one possible option, though the challenges of administering and monitoring would need to be addressed.

Further, the government must do more to promote infrastructure sharing. Both passive (towers, rental places) as well as active (antenna, feeder cable, Node B, radio access network and transmission systems) infrastructure sharing amongst service providers are allowed (DoT, March 2008). However, in order to effectively use the radio access network infrastructure, incentives in the form of lower regulatory levies for sharing need to be provided to improve adoption. This in turn will improve the efficiency of the radio access networks. An article on the effective use of mobile infrastructure is given in Box 4.2 (Prasad and Sridhar, 19 May 2008).

Spectrum Pricing

The government has chosen to provide the UAS licence for upfront fixed fees along with a revenue-sharing agreement. The revenue share of the government is increased by a percentage point on every fresh allocation of spectrum. Subscriber-based norms are used to determine eligibility for fresh spectrum. The fixed fee was fixed to the fee paid by the fourth cellular operator in the auction of 2001. A benchmarked price is appropriate only if the object being sold (in this case, a licence) is the same, and the market conditions are similar. The fee paid by the fourth operator in the auction process was specifically for a cellular mobile telephone service (CMTS) licence along with start-up spectrum. The UAS allows the licensee to provide access services using non-spectrum-related technologies such as wire-line service as well. As per the guidelines of the UAS licence (DoT, 2005):

The unified access services cover collection, carriage, transmission and delivery of voice and/or non-voice MESSAGES over LICENSEE's network in the

Box 4.2   Are Operators Hoarding Spectrum?

Mobile operators deploy BTSs at cellular towers to facilitate wireless communication between mobile phones and the mobile network. The communication takes place using the radio spectrum allotted to the operators. Among other factors, the amount of spectrum the operator has and the number of BTSs deployed directly impact the subscriber base the operator can support with a defined quality of service. The operator maximizes profit by using a judicious combination of the two main inputs—spectrum and BTS. The profit-maximizing condition requires that the marginal product earned from the last Rupee spent on each input be the same. Upward revisions of subscriber norms for additional spectrum allocation and the exclusion of incumbents from the recent allocation process for 2G spectrum seem to indicate that the government is of the view that the incumbents are hoarding spectrum.

Our analysis of the data on subscriber base, spectrum allotted and BTSs deployed by the Indian operators from 2001 to 2007 indicates that the marginal product per Rupee of spectrum is greater than that of BTS. This seems to imply that under the current market conditions operators would like to substitute spectrum for BTS in order to minimize cost of service and increase profits. Therefore rather than hoarding spectrum, which means holding spectrum beyond the amount required to maximize near-term profits, operators are unable to maximize profits because of a shortage of spectrum.

However, the apparent scarcity could also be because of the following reasons.

One: Spectrum is under-priced relative to BTS. The Planning Commission had expressed its concern about the low price level of spectrum. The original price revealed in the spectrum auction of 1995 was tempered by the government's bailout of mobile operators in 1999. After the multi-stage auction for the fourth operator licence in 2001, start up spectrum was allotted to new entrants using the 2001 price levels which were low. The spectrum usage charges based on revenue share are not rationalized to reflect the true value of this scarce resource.

Two: The operators are not actively using methodologies such as cell sectoring, deployment of directional and smart antennas, utilization of Adaptive Multi-rate (AMR) encoding techniques to increase the subscriber base and hence traffic supported per cell site. These involve investment and change in network configurations. The operators are reluctant to implement these due to the possible availability of additional spectrum at very low prices (if luck prevails!). Moreover, deploying more BTSs makes better sense than making each BTS more efficient. Though recent policies and the market environment have promoted sharing of active and passive infrastructure, these changes are at a very nascent stage.

We examined the hypothesis that correct pricing of spectrum and more efficient usage of radio access network would eliminate spectrum scarcity using prices and productivity of spectrum and BTS. We find that at current levels of spectrum holding, which are far higher than the subscriber-based norms prescribed by the DoT in January 2008, the combined percentage increase of the price of spectrum relative to BTS and the efficiency of BTS relative to spectrum would need to be 100 per cent in order to eliminate scarcity of spectrum. If we peg the usage of spectrum as per the subscriber-based norms, then the combined percentage increase required to eliminate the phenomenon of spectrum scarcity would be to the order of 300 per cent. This large increase of the order of 300 per cent, or even of 100 per cent indicates with high probability that spectrum is in short supply both at present levels of usage, and at the levels envisaged in the subscriber-based norms.

Given this conclusion many recent policy decisions seem untenable. The decision to completely exclude incumbents from the recent allocation of 2G spectrum, the low fixed price determined for the allocation of start-up spectrum to new entrants, the government's inability to formulate policies to facilitate optimal sharing of mobile infrastructure amongst the operators, are lacuna that need to be addressed.

Contrary to popular perception, incumbents still need more spectrum.

designated SERVICE AREA and include provision of all types of access services. The access services include but are not limited to wire-line and/or wireless service including full mobility, limited mobility as defined in clause 12 (c) (i) and fixed wireless access.

The UAS licence includes not only wireless services but also other associated services, and therefore the price paid for it should in general be higher than that paid for CMTS licence. Further, market conditions in 2003, and even more starkly, in 2007 were quite different from those in 2001. Let alone factoring in the increased value of spectrum in the booming telecom sector, the government did not even factor in inflation to arrive at the 2003 and 2007 prices.

Inappropriate benchmarking has resulted in under-pricing of spectrum. Under-pricing spectrum leads to a tendency of hoarding and therefore should be avoided.

The confusion regarding the CMTS and UAS licenses has cut both ways since UAS licensees providing CDMA service and applying for a GSM licence in 2007 had to pay exactly the same amount that they paid when they were issued the CDMA-UAS licence. In fact, as per the above argument, only the charge for GSM spectrum should have been levied.

Though UAS implicitly separated licensing from spectrum, licence fee needs to be separated from spectrum fee for pricing both licences and spectrum appropriately (in case a fixed fee method is chosen).

Regulatory Certainty

Telecom licenses should balance regulatory certainty with the flexibility necessary to address future changes in technology, market structure and government policy (Intven *et al.*, 2000). The fixed fee is paid by the UAS licensees without any firm guarantee on the date of allocation. It therefore involves a promise to allocate spectrum at an uncertain point in the future. Further, the subscriber-based norms represent a very complicated future contract in spectrum with an additional clause of the seller having the right to renegotiate the terms by strengthening the subscriber-based norms. This contract is very non-transparent and difficult to convert into monetary terms for the purpose of decision-making by the involved parties.

Moreover, there is no injunction on the regulator for an orderly step by step change of subscriber-based norms in response to technological changes. Thus the government decision on strengthening of subscriber-based norms by an order of six to fifteen times in effect increased the price of spectrum from nil to infinity (since the incumbents became ineligible for fresh allocation). This does not represent a very conducive environment for business decision-making.

Technology Neutrality

The introduction of CDMA-based mobile service during 2001–3 was clouded by the legal battles between the incumbent GSM operators and the BTS operators. Given their defensive position and the prevailing paradigm of maximal usage, the CDMA mobile service providers were allotted a lower start-up spectrum block (2.5 MHz as compared to 4.4 MHz in GSM) as they were considered more efficient. Parity was established between the two technologies in terms of the number of subscribers each could reach with the spectrum allotted to them.

The definition of parity is tantamount to handicapping a more efficient player by restricting access to a necessary resource. In an environment where maximal usage is no longer relevant, the correct definition of parity is 'equal access to resources' to be used as inputs. Such an approach would create a truly level playing field in which the respective technologies would be able to compete. Promoting such competition would also economize on the use of spectrum. The only qualification to this conception of parity could be the threat of monopoly power. However, if anything, the strengthening of CDMA service would increase competition in an industry that is currently largely dominated by the GSM players.

As per the TRAI (2005) both types of operators should have the same amount of start-up spectrum and be held to the same subscriber-based norms.

Level Playing Field between Government and Private Operators

As is pointed out in the TRAI (2007), the ratio between the actual subscriber base and the number of subscribers that should be serviced as per the subscriber-based norms is much larger for private operators compared to government operators, indicating that private operators are having to stretch spectrum much more than the government operators. In some LSAs this ratio is even less than one for the two government operators, indicating that they were given spectrum despite not meeting the subscriber-based norms. Even when both the government and private operators met the subscriber-based norms, government operators were given priority in the allocation of spectrum. This places them at an advantage in an environment where spectrum is scarce and subscriber-based norms change rapidly. In a highly competitive mobile services industry, such practices do not indicate the presence of a level playing field. If subscriber-based norms are to continue, then the additional spectrum should be allotted to eligible operators on a first-come-first-serve basis.

However, the policy announcement is silent on the road ahead after the present allocation. One hopes that the mistakes of 2G are not repeated in 3G. A clear roadmap of spectrum availability, use of the auction mechanism for allocation, rationalization of the subscriber-based norms to serve as a low hurdle of eligibility to participate (as opposed to requiring maximal usage) would be appropriate. Prasad and Sridhar (9 June 2007) give a detailed analysis of the expected reserve price in comparison to the international prices. The article in Box 4.3 compares spectrum with real estate and describes measures to be taken while allocating spectrum (Sridhar, 29 September 2006).

Need for Consolidation

Given the low spectrum holding per operator and the evidence of allocative inefficiency in the industry, policy should be formulated to enable the consolidation of the market and hence of spectrum through merger of licenses, transfer of allotted spectrum from one licensee to another, or sharing of spectrum amongst licensees. At the present level of spectrum fragmentation it is unlikely that the target of 1 billion

Box 4.3   To Allocate Spectrum, Study Real Estate

Radio frequency spectrum being a scarce resource, its management is important and should serve at least the following purposes: a) granting of exclusive rights to spectrum (b) ensuring its efficient use and c) promoting competition in services. A key issue in managing the above optimally is to define adequate property rights for the spectrum.

It is useful to compare spectrum property rights to the property rights on real estate. In a competitive market, spectrum is bought at a price by operators much like buying real estate, using methods such as auction, beauty contests and lotteries. The TRAI's recommendation on ascending auction with a reserve price is a transparent mechanism for allocating the scarce resource. It determines a reliable market price through iterative price discovery by the bidders and prevents the classical winner's curse problem. Ascending auction allows the learning needed to identify the intersection of supply, demand and hence the market price. A reserve price is required to minimize collusive behaviour. However, the recommended reserve price (of about Rs 1,400 crores for pan-India presence) seems to have been fixed by the intentions of one of the operators and more by the government's intentions to fleece the operators. The DoT shall consider lowering the reserve price substantially so that the correct market price is determined through the auction procedure.

The property rights of the real estate owner can also include certain restrictions on usage, for example various building regulations, in order to avoid interference with neighbouring properties and general public interests. In order to avoid interference issues, the WPC wing of the DoT has developed the National Frequency Allocation Plan (NFAP) and specified the usages of various frequency bands for various applications. Technology innovations in telecom continue to give birth to new applications that interfere with legacy applications. For example, WiMAX technology that operates in the 2.5 GHz and 3.5 GHz bands interferes with the band currently being used for Indian National Satellite (INSAT) communications of the Department of Space (DoS). Mobile TV applications that use the 700 MHz band interfere with traditional TV broadcasting. It is important to revisit and examine these interference issues critically from time to time. The TRAI has done a tremendous job of identifying the above various bands for 3G and BWA applications. Specifically, allocation of the 450 MHz band will give impetus for rural deployment. Allocation of additional spectrum in the 800 MHz band and recommendations on re-farming of the 1900 MHz band certainly level the spectrum playing field between the GSM and CDMA operators.

In real estate, if a certain piece of land is needed for a public purpose, for example building of a public infrastructure, it may be possible to expropriate the property. To give an example of telecom, in India, the better part of the spectrum in the 1800 MHz band used for GSM-based cellular mobile services in the country is being used by the Department of Defence (DoD).

The government has recently been making efforts to release up to 45 MHz spectrum from the Defence in the interest of the growing mobile services in the country. The TRAI's recommendations for the release of 100 MHz in the 3.4–3.6 GHz band, currently used by the DoS for BWA applications is a step in the right direction.

Similarly, unused land resources may be confiscated and allotted to users keeping in mind public interest. The TRAI's recommendations on rollout obligations and penalty for not meeting them including cancellation of spectrum assignment will definitely prevent hoarding of the scarce resource.

There are various stakeholders involved in the spectrum game: DoD and DoS who are large users of various spectrum bands; Ministry of Finance (MoF) which looks at spectrum allocation from the economic point of view to ensure that the existing supply of radio spectrum is used in a way that maximizes the economic value; and DoT that uses the technical approach of optimally allocating spectrum resources to promote competition and minimize the interference between various applications. The TRAI's recommendations on the setting up of the National Frequency Management Board (NFMB) will definitely alleviate coordination problems amongst various stakeholders.

Clarity on spectrum allocation is needed for sustaining the growth and competition in mobile services. The NFMB, when set up should strive for a detailed frequency plan similar to the Interactive Frequency Plan published by the Danish Telecom Agency, incorporating the spectrum demands of the current and emerging technologies and applications. It is a good beginning by the regulator (of course subject to acceptance by the DoT—the policymaker) for a transparent long-term view of spectrum management.

subscribers can be serviced with mobile voice and data services. While allowing consolidation, restrictions on significant market power should be achieved by increasing contestability[3] in the market.

In the early stages of the evolution of the mobile industry, almost all countries including India followed a command and control policy in respect of spectrum allocation and management. This included identifying the 'best use' of the band, choosing users (for example, based

---

[3] William Baumol in a classic article in the *American Economic Review* way back in 1982, espoused 'contestability' as a broader benchmark of competition than market power as measured by the HHI (Baumol, 1982). A 'contestable' market is one in which entry and exit of firms are free. In this case the incumbent(s) face a real threat of rapid entry and exit of firms if they make supernormal profits. Hence a contestable market never offers more than a normal rate of profit and is characterized by the absence of any sort of inefficiency in production.

on first-come-first-served or beauty contest method), setting prices, and achieving efficiencies/consumer benefits by intervention (for example, rollout obligations, 'use it or lose it' conditions). The command and control approach is useful for achieving early rollout and rapid growth. However, there are attendant risks of regulatory failures including spectrum getting 'stuck' in lower value uses, and absence of mechanisms to put under-utilized spectrum to a better use. Market mechanisms such as auctioning of spectrum should be explored as an alternative, since they allow price revelation, and allocate spectrum to those who are best able to use it in the highest value applications.

Property Rights and Change of Use

Procedures for trading and changing use or other parameters of the licence vary significantly across countries. For example, in Australia and New Zealand, the property rights are broadly defined according to technical or core parameters that set the maximum level of emissions. If this level is exceeded, the affected licensee has a right to force the licensee that causes interference to take measures to reduce its emissions. However, the licensee is free to act as it wishes including modifying the nature of the services offered, as long as there is adherence to interference guidelines (Crocioni, 2009).

In India, the government has restricted the use of spectrum allotted in the 900 and 1800 MHz bands exclusively to provide 2G services. The provision of 3G services is restricted to the 2100 MHz band. However, 3G technologies such as CDMA-evolution data optimized (EVDO), and WCDMA can be technically provided in other bands (for example, 900, 1800, and 1900 MHz) as well. These restrictions on use hinder trading. The existing commercial users of spectrum have very little incentive to sell excess or unused spectrum if the buyer will use its acquired spectrum to provide a service that is currently provided by the seller (for example, 2G or 3G). Consequently, the number of participants in a spectrum market may be very low. Such market 'thinness' decreases the likelihood that a trade will take place (Bykowski, 2003). For example, in Australia, the lower activity levels in the trading market are attributed to the lack of portability and the nature of property rights specified in the licence (Xavier and Ypsilanti, 2006).

So the policymaker and regulator should make spectrum available in a technology-agnostic manner. Hence change of use of a spectrum band should be allowed after trading so that the buyer deploys the acquired spectrum block in its most effective use, be it 2G or 3G. Such a spectrum-trading market would allow firms to choose technologies based on market conditions rather than on standards dictated by the government. This will also allow a *de facto* standard to emerge, if appropriate.

Mergers and Acquisitions

In European countries, consolidation has taken place through mergers and acquisitions (M and A). However, in India, the following clauses in the mergers and acquisitions' guidelines limit mergers and acquisitions (DoT, 2008a):

1. The market share of the merged entity in the relevant market shall not be greater than 40 per cent either in terms of subscriber base or in terms of adjusted gross revenue. The subscriber base and the associated market share threshold is calculated separately for wireless and wire-line subscribers.
2. No M and A activity shall be allowed if the number of UAS/CMTS access service providers reduces below four in the relevant market consequent upon the M and A activity under consideration.
3. The merged entity is subject to the SBC and has a period of three months after the merger to fulfil the SBC with respect to the aggregate spectrum jointly held.
4. Any permission for merger will be given after the passage of a minimum of three years from the effective date of the licence.

In future it is preferable that the government loosens these norms and allows the Competition Commission of India, the regulatory body set up by the Indian government to assess competition issues, to address the issue of significant market power, if any.

Changes in Sharing Arrangements

Trunking efficiency gains can also accrue through sharing of spectrum. Sharing of spectrum can be beneficial (i) when there are pockets in an LSA where an operator has less spectrum, (ii) amongst operators having non-uniform and complementary subscriber bases in different parts (say, urban and rural) of an LSA (iii) when two or three operators with less spectrum (say 2 × 4.4 MHz each) want to set up a common network with pooled spectrum.

Sharing of 2G spectrum amongst UAS/CMTS licensees will be facilitated if the annual spectrum usage charges are made uniform for all bands irrespective of amount of spectrum held, as recommended in

Sridhar and Prasad (2011).[4] Sharing should be permitted on payment of 'sharing charges' to the government for the quantity of spectrum shared, in the same manner and of like amount as applicable in case of transfer or merger of the spectrum.

When two operators share spectrum, sharing charges shall be levied on the smaller of the two spectrum blocks being shared. In case three operators share spectrum, sharing charges shall be levied on the smaller two spectrum blocks being shared.

Since spectrum-sharing arrangements may sometimes stop, the formulated policy may also provide for retention of sharing charges only to the extent levied for the actual period (part of the year will be taken as full year) of the sharing on a pro-rata basis, and refund of the difference. In case of subsequent sale or merger of the spectrum, transfer charges or merger charges as the case will be payable, pro-rata on the balance period of the spectrum assignment.

In case of sharing of spectrum, each licensee should have the benefit of the aggregate shared spectrum. For the purpose of assessing the total 2G spectrum holding of a UAS/CMTS licensee, the total shared spectrum should be counted in the hands of each licensee. In case one of the licensees sharing the spectrum has already fulfilled the rollout obligations, there should be no further penalties on any of the licensees sharing the spectrum.[5] In the case where none of the licensees has fulfilled the rollout obligations, penalties for unfulfilled rollout obligations need to be applicable on each licensee separately. In case of sharing it will be necessary to prescribe responsibility related to frequency, power limits and interference, jointly and severally for compliance of licence conditions of the entire shared spectrum. Details on spectrum trading as applicable in the Indian context are explained in Sridhar and Prasad (2011).

---

[4] With increasing spectrum usage charges, the government would need to decide the usage charges to be levied on the two parties in a sharing arrangement. There are three options: continue with the charges prior to sharing; charge both the parties the higher or lower of the two rates fixed prior to sharing; apply the rate applicable to the combined spectrum block on the assumption that each party uses the whole block. Each of these options comes with its own challenges.

[5] Note that while rollout obligations will not apply to new licensees, liabilities that have already been incurred on this account cannot be waived. However, in case of sharing, they can be modified as suggested.

Apart from the above, complementary measures such as spectrum refarming should be explored. The US and many European countries are advocating spectrum refarming where spectrum in ultra-high frequency traditionally used for terrestrial TV broadcasting is being made available for commercial mobile services (Jain and Jain 2007). On 12 June 2009, all the US terrestrial broadcasting stations switched off their analogue transmission and turned on digital transmission, thus freeing up spectrum for mobile and emergency services. These initiatives have enabled several countries (for example, the US, UK, Korea, France) to reap the benefits through additional spectrum (108 MHz in the US, 128 MHz in the UK, 72 MHz in France, and 54 MHz in Korea) being made available for commercial purposes. Moreover, many European countries have started reforming the 900 MHz band traditionally used for GSM 2G services for UMTS 3G services.

Further, spectrum sharing should be evaluated as an option that allows spectrum to be used most efficiently. Technologies such as Dynamic Spectrum Access (DSA) promise to increase spectrum sharing among competing service providers and thus help overcome the lack of availability of spectrum (Chapin and Lehr, 2007). One of the advantages of using DSA is that it allows real-time trading of spectrum access rights and use of high-power transmissions at times when the primary users of a frequency band are inactive.

A comprehensive list of measures for improving the efficiencies of the spectrum and the industry are presented in Sridhar and Prasad (2011).

* * *

The spectrum policy in India while having been very successful in nurturing the growth of the industry suffers from lack of a long-term vision and absence of a holistic perspective that considers all the relevant factors before making policy decisions. The trajectory of the spectrum policy in India has been marked by many flip-flops: on subscriber-based norms, spectrum pricing, 3G policy, and competing technologies. The resulting uncertainty is harmful for the industry. A clearly defined, consistent policy is the need of the hour.

The Indian telecom industry is facing legacy issues related to the allocation of spectrum. It can address these issues by allowing the operation of secondary markets. The introduction of trading/sharing/ mergers needs to be accompanied by several policy measures designed to allow the process of consolidation to take place in an equitable, efficient, and transparent manner. Despite the legacy inhibitors to trading/sharing, a new policy paradigm has several advantages. It will reduce industry costs, promote innovation and trigger diffusion of mobile services (Valletti, 2001).

Spectrum trading could also make innovative activities less risky. If a firm decides to launch a new service based on a new technology or a product (compatible or incompatible with existing ones), it can buy the spectrum it needs and market its products. If the product is successful, the firm may want to expand by buying additional spectrum. If the product does not prove popular with consumers, the firm could resell the spectrum rather quickly and reduce the losses from potential failures (Valletti, 2001). Sridhar and Prasad (2011) in their recent study indicate the methods by which spectrum consolidation can happen in India and the prerequisites for a successful secondary market.

5

National and International Long-distance Services
Does Distance Matter?

National Long-distance Service

National Long-distance (NLD) service refers to the carriage of switched bearer telecom service over a long distance. The NLD service was provided by the DoT of the Government of India until 2001 as a monopoly service. As per NTP 1999, the Indian government decided to liberalize the provisioning of NLD services by inviting private firms to enter into providing the service in 2001. The licence guidelines (DoT, 2001) state that 'NLD operator (NLDO) licensee will have a right to carry inter-LSA traffic. Intra-LSA traffic can be carried by NLDO with mutual agreements with fixed/mobile service provider in accordance with their mutually agreed terms'.

While NLD was opened up in 2001, the access licenses (basic and cellular mobile) were opened up for competition as early as 1994. This was a marked departure from the practice in many other countries. For example, in the US, there were many operators providing services in 1984 when the Modified Judgement created the long-distance arm of AT&T. However, only in 1996, after the implementation of the Telecom Act, was the natural monopoly in basic telecom service broken up. Unlike access service, laying down backhaul networks is not very expensive, considering the availability of the terabit-capable high-capacity optic fibre. Due to traffic aggregation, trunking efficiency is easier to attain in an NLD network compared to access network, resulting in lesser cost per unit of information transfer. Hence most countries introduced competition in NLD prior to access service provisioning. An example of a NLD network is given in Figure 5.1. The NLDO sets up a switching and transmission centre, referred to as the PoP at the Long-distance Charging Area (LDCA) level to provide on-demand inter-LSA long-distance service.

The National Long-distance Licence Process

In 2001, the government announced NLD guidelines (DoT, 2001). An entry fee of Rs 100 crores was fixed for the award of the licence for a period of twenty years. Apart from the entry fee, the licensee had to meet certain minimum requirements such as paid-up equity capital (of Rs 250 crores), combined net worth (of about Rs 2500 crores) and experience in telecom.

The twenty-one different LSAs as specified in the BTO licence were covered under the licence. Each LSA was divided into different LDCAs. The licensees had to submit bank guarantees to ensure their commitment towards their rollout obligations, which are given in Table 5.1. In the event that the rollout obligation could not be fulfilled, then the bank guarantees would have to be forfeited by the licensee. Apart from the entry fee, an annual licence fee of 15 per cent (including 5 per cent universal service levy) of the AGR was levied on the NLDOs. There would be no cap on the number of operators.

Bharti and Reliance were given the licence after payment of the stipulated entry fee. The government-owned VSNL whose monopoly in International Long-distance (ILD) was broken up prematurely in March 2002, two years ahead of planned divestment (details of which are available in the subsequent section), was awarded NLD licence as part of the compensation package along with the required entry fee and annual licence fee for five years.

Table 5.1 Minimum Rollout Obligations for National Long-distance Operators for Establishing Point of Presence

Phase	Time Period (in years, from the Effective Date of Licence)	Cumulative National Coverage at the LDCA Level where PoP has to be Established	Cumulative Percentage of Uneconomic and Remote Areas
I	2	15	2
II	3	40	4
III	4	80	7
IV	5–7	100	All

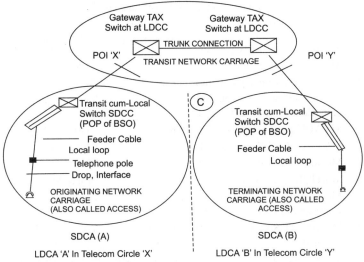

Figure 5.1   Interconnection Network

Source: TRAI, December (2001)

In December 2005, with a view to increasing competition, the government reduced the entry fee to Rs 2.5 crores (DoT, December 2005). The annual licence fee was also reduced from 15 per cent to 6 per cent, inclusive of universal service levy of 5 per cent. The annual licence fee paid by NLD operators to the government in the first quarter of 2008–9 (April–June 2008) was about Rs 1.4 billion and contributed to about 6 per cent of the licence fee collected for all telecom services in the country.

Table 5.2 gives the list of NLD licensees. A look at the table indicates that foreign operators such as Cable & Wireless, AT&T and British Telecom have got the NLD licence. These operators currently provide business and enterprise solutions to large software and BPO companies in India.

National Long-distance Market

The NLD revenue almost doubled from 2002 until 2008. Growth of revenue since the market was liberalized in 2001 is given in Table 5.3 (*Voice and Data*, 2008).

Table 5.3 indicates doubling of the year-to-year growth in 2005-6. This was due to reduction of entry fee and the subsequent entry of many operators, thus increasing competition. The increase in competition led

Table 5.2   List of National Long-distance Licence Holders

S.No.	Name of the Licensee	Licence Signed on
1.	M/s Bharat Sanchar Nigam Ltd.	Incumbent Operator
2.	M/s Bharti Airtel Limited	29 November 2001
3.	M/s Reliance Communications Limited	28 January 2002
4.	M/s Videsh Sanchar Nigam Ltd. (Tata Communications Limited)	8 February 2002
5.	M/s Mahanagar Telephone Nigam Ltd.	10 May 2006
6.	M/s Power Grid Corpn. of India Ltd.	5 July 2006
7.	M/s RailTel Corpn. Of India Ltd.	7 July 2006
8.	M/s HCL Infinet Ltd.	11 July 2006
9.	M/s i2i Enterprises Ltd. (BT Global Communications India Pvt. Ltd.)	11 July 2006
10.	M/s Tulip IT Services Ltd. (M/s Tulip Telecom Limited)	8 August 2006
11.	M/s Shippingstop Dot Com (India) Pvt. Ltd. (M/s Loop Telecom Pvt. Limited)	18 September 2006
12.	M/s AT&T Global Network Services India Pvt. Ltd.	9 October 2006
13.	M/s Vodafone Essar South Ltd.	10 November 2006
14.	M/s Sify Communications Ltd.	21 November 2006
15.	M/s Idea Cellular Ltd.	23 November 2006
16.	M/s Dishnet Wireless Ltd.	13 December 2006
17.	M/s BT Telecom India Pvt. Ltd.	20 February 2007
18.	M/s Tata Teleservices Ltd.	30 July 2007
19.	M/s Spice Communications Ltd.	8 August 2007
20.	M/s Oil India Limited	27 December 2007
21.	M/s Verizon Communications India Private Limited	3 January 2008
22.	M/s Cable & Wireless Networks India Private Limited	15 February 2008
23.	M/s Equant Network Services India Private Limited	20 June 2008
24.	M/s Swan Connect Communications Private Limited	12 August 2008
25.	M/s Citicom Networks Private Limited	3 October 2008
26.	M/s Swan Telecom Private Limited	6 October 2008
27.	M/s SingTel Global (India) Private Limited	5 March 2009
28.	M/s Datacom Solutions Private Limited	18 March 2009
29.	M/s Unitech Long Distance Communication Services Limited	28 April 2009

Source: DoT, July 2009a.

Table 5.3   Revenue of National Long-distance Market

Year Ending	Revenue in Rs Million	Y-Y Growth	BSNL Market Share
March 2002	59,700		92%
March 2003	59,880	0.30%	
March 2004	51,410	−14.14%	88%
March 2005	62,610	21.79%	81%
March 2006	90,150	43.99%	
March 2007	71,860	−20.29%	
March 2008	97,320	35.43%	30%
March 2009	144,320	48.29%	
March 2010	164,000	13.64%	22%

Source: Voice and Data, July 2010a.

to sharp decline in prices. However, there was a time lag for demand to increase proportionately. This led to a huge decline in revenue and negative growth in the subsequent year. The competition and low prices propped up demand, which is indicated by the steady increase in revenue from 2007 onwards.

An interesting observation is that the share of the erstwhile monopoly government operator (BSNL) is continuously declining, more so in recent years. Table 5.4 illustrates how the private operators have been able to grab the market share from the incumbent.

Most major access providers (basic and cellular mobile) such as Bharti Airtel, Reliance, Vodafone, and Idea provide NLD services. One exception in Table 5.4 is Tata Telecommunications (erstwhile VSNL) which does not have the licence to provide access services. Tata Telecommunications provides NLD services for access operators that operate in distant, non-contiguous licence areas. For example, the government operator MTNL operates only in Delhi and Mumbai for whom NLD services were provided by Tata Communications.

INTERCONNECTION AND EQUAL ACCESS

Interconnection is defined as one carrier using the infrastructure owned by another carrier in the same industry (Shy, 2001). Interconnections are mostly observed in industries where the fixed and sunk costs associated with the infrastructure are significant compared to the cost of carrying or transmitting one unit of output over these types of infrastructure. Hence it

Table 5.4   Growth and Market Share of National Long-distance Operators

NLDO	Revenue (in Million of Rs)		Y-Y Growth	Market Share
	Ending March 2009	Ending March 2010		
BSNL	31,590	36,000	13.96%	22%
Bharti Airtel	49,050	48,000	-2.14%	29%
Reliance	27,290	21,000	-23.05%	13%
Tata Communications	10,500	10,450	-0.48%	6%
Vodafone	13,280	19,000	43.07%	12%
Idea	7,130	10,000	40.25%	6%
Others	5,480	19,550	256.75%	12%
Total	144,320	164,000	13.64%	

Source: Voice and Data, July 2010a.

is of importance in the telecom industry, especially in the access networks. The calls originating in an access network have to be carried over the NLD network of the NLDOs and hence the importance of interconnection. The NLDO needs to access the often dominant access provider (basic or cellular mobile) to reach customers. The access infrastructure is sometimes referred to as the 'essential facility' or the 'bottleneck' facility (Shy, 2001). The access facility, especially the basic wire-line facility, is monopolized as explained in previous sections due to large economies of scale, hence acts as bottleneck facility. However, the wireless mobile services are often competitive, hence do not act as bottlenecks but are essential facilities for originating or terminating any long-distance calls. Under these conditions, the telecom regulator normally intervenes to create proper conditions for access to essential facilities by the NLDOs and avoid bypasses. The DoT (2001) also mentioned that the NLDO licensee shall operate and maintain the licensed network conforming to the quality of service standards to be mutually agreed between the service providers in respect of network-network interface. This was to ensure that the NLD network seamlessly interconnects with the different access networks at the PoPs.

With the introduction of competition, the NLD segment in India entered in to a multi-service, multi-operator scenario. Interconnection of different networks, though complex, is critical for the success of open competition. In most of the countries, 'equal access' to the network of the access providers is mandated to be provided to all the NLD operators so that the above problem is solved to some extent.

INTER-CARRIER COMPENSATION

Prior to liberalization, the NLD service was provided by the DoT of the Indian government. The access service was primarily through the monopoly wire-line service also offered by the DoT until (i) the competitive local exchange carriers came into existence in 1998 and (ii) the cellular mobile service was launched in 1995. During this monopoly period, NLD services were considered a luxury and were charged at a premium compared to local access services. The NLD call prices were kept high by the monopoly carrier so that the charges could be used to cross-subsidize local call charges to reflect the policy of the government to provide local access service at affordable prices to the masses of the country. The NLD charges were (i) distance-sensitive (that is, greater the distance higher the call charges/ min) and (ii) time-sensitive (peak and off-peak pricing to optimize the constrained bandwidth on the NLD network). When competitive mobile and wire-line access services were introduced, there was a need to introduce interconnection regulation so that (i) the different access providers could connect to the NLD network of the incumbent and (ii) the carriage fees levied by the NLD carrier would be economically attractive for the new access providers to connect to the NLD carrier's network. Further, when competition in the NLD market was introduced in 2001, there was a need to introduce inter-carrier compensation for successfully terminating the NLD calls of the new entrant with subscribers of the erstwhile monopoly operator. We illustrate here the two cases of inter-carrier compensation policies that were to be addressed by the regulator and the government. Figure 5.2 illustrates the one-way access of a NLD carrier.

A call carried over a new NLDO network bypasses that of the incumbent NLDO. If the terminating incumbent is a monopoly (which often is the case in wire-line basic services), there is no incentive for the incumbent to terminate the call. Hence the new NLDO entrant is required to compensate the incumbent with an 'access charge' which is just sufficient to sustain the interest of the incumbent for termination. In this example, a 'one-way access charge' is paid by the new entrant to the incumbent.

The access charge that the NLDO needs to pay to the incumbent for terminating or originating the call is determined as follows (Shy, 2001):

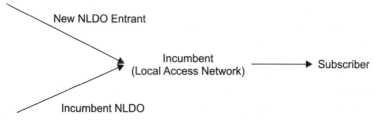

Figure 5.2 Interconnection between National Long-distance Operator and Local Access Provider

1. The entrant pays for the marginal cost generated by having the incumbent carry the call to its final destination, plus its share of the fixed cost according to its relative use of the local switch. OR
2. The entrant's access price simply compensates the incumbent for the incumbent's loss of profit due to transition of long-distance customers to the entrants.

The above is often referred to as 'inter-carrier compensation' and is often specified in the interconnection regulation. The situation is complex if the originating and terminating access providers and the type of access technology used are different. For example, take the case of Bharti Airtel mobile subscriber A who makes a long-distance call to a BSNL landline subscriber B. The Airtel mobile access network picks up the call made by A, takes it to the point of interconnection and hands over the call to the BSNL long-distance network which in turn carries the call to the switch nearest to B and terminates it on BSNL's landline network. If all Airtel mobile customers start making such calls, soon BSNL's long-distance network would get congested.

Since the inter-LSA calls are carried by the NLDOs, it should be mandatory for BTOs, CMSPs and UASPs to provide interconnection to NLDOs so that the subscribers could have a free choice to make inter-LSA calls through NLDOs. The DoT (2001) clearly mentions the above condition, and extends this interconnection clause to Cable TV operators as well. Apart from providing interconnection, each access provider must respond to such an increase in inter-network calls by investing in switching and transport capacity as per the interconnect regulation.

Assume a scenario in which long-distance calls originate from a basic/ mobile access provider such as Bharti Airtel. If the majority of calls are placed by Bharti Airtel subscribers to BSNL subscribers, BSNL incurs extra network costs of its own that are attributable to calls originated on Bharti Airtel's network by Bharti Airtel's subscribers, with whom BSNL may have no contractual relationship of any kind. However, BSNL must purchase high-capacity switching and transport network elements to accommodate the flood of incoming calls. This example illustrates a complex scenario in which the important question is who should pay for the additional network costs that a carrier incurs because of its duty to terminate calls originated on other telecom networks by callers with whom the carrier may have no direct relationship. The most widely used solution to this problem requires that the calling party's carrier, whose customers originate the calls, that cause these extra costs, compensate the terminating carrier that incurs the costs. This access charge is passed on to the customer by the calling party's carrier.

On 1 May 2003, the Telecommunication IUC regulation notified by the TRAI came into existence and was further modified in October 2003 (TRAI, 2003). The TRAI (2003) clearly defines IUC as 'the charges payable by one service provider to one or more service providers for usage of the network elements for origination, transit or termination of the calls'. Since the CPP scheme is present in both mobile and landline services, the IUC regulation specifies the IUC which can be charged by the calling subscriber's carrier for billing purposes. The calling party carrier then passes on the associated IUC to the NLDO as well as the terminating carrier. The TRAI (2003) specified only termination and carriage charges, which are given in Table 5.5.

In the IUC regulation, the originating charges were in forbearance due to the highly competitive mobile services. The originating service provider shall retain origination charges from the residual after payment

Table 5.5 Interconnection Usage Charges for Termination and Carriage (as of 2003)

Termination Charge for Fixed Line, Cellular Mobile (in Rs/ Min)	NLDO Carriage Charges (in Rs/Min)			
	Distance <= 50 km	50 km < Distance <= 200 km	200 km < Distance <= 500 km	Distance > 500 km
0.30	0.20	0.65	0.90	1.10

of the charges for carriage, termination and access deficit (as explained in a previous chapter). The above charges were calculated based on the historical capital and operational expenditures of the government operator appropriately amortized and after consultation with all stakeholders concerned. Through the IUC Regulation Order dated 23 February 2006, the distance-based carriage charges were revised to a uniform carriage charge of Re 0.65 (TRAI, May 2008).

In June 2008, the DoT asked the TRAI to re-look at the IUC, especially the termination charges which were perceived to have come down due to technological advances and competition. In its March 2009 recommendations, the TRAI revised the IUC as follows:

1. Termination charge for all types of domestic calls (namely fixed to fixed, fixed to mobile, mobile to fixed and mobile to mobile) has been reduced to 20 paise per minute from 30 paise per minute.
2. Termination charge for incoming international calls would be 40 paise per minute.
3. Ceiling on carriage of domestic long-distance calls retained at 65 paise per minute.
4. Origination charge has not been specified as it would be residual from tariff after payment of other charges.

CARRIER SELECTION

Carrier selection is defined as a method to provide control to the calling part as to who carries the call. In a multi-operator environment, carrier selection is a prerequisite for competition benefits to reach the end-subscriber. Carrier selection allows the subscriber to specify who should carry each portion of the call and this may be achieved primarily either (i) through prefixing the carrier-allotted code or (ii) through carrier pre-selection (TRAI, May 2008). With as many as twenty-three or more operators, carrier selection assumes importance as it provides competitive players with the opportunity to grow and consumers to derive benefits from innovative tariff plans. Option (ii) can be exercised by the subscriber in coordination with the local access provider. Unless stated otherwise, the subscriber's NLD voice traffic will be carried through the selected NLD carrier, also referred to as 'primary long-distance operator' as in the US and other countries. Option (i) provides the subscribers with a 'secondary long-distance operator' of choice so that call-by-call selection of the NLD carrier can be made. Healthy competition amongst the NLD carriers is possible only when subscribers are able to select carriers of their choice for their NLD calls.

Clause 2.2 (d) of the NLD licence states that NLD providers can access the subscribers directly only for the provision of leased circuits/closed user group services. Hence the NLDOs cannot reach the residential or business subscribers directly for the provisioning of public NLD voice service. The absence of carrier selection aggravates this problem by preventing end-subscribers from exercising their choice of the NLD network. The long-distance minutes were bundled by the access providers and sold along with local access minutes. Hence even though many operators were providing NLD service in the country, the competition was not visible.

Sector regulator, the TRAI, carried out a consultation process way back in 2001 and subsequently gave directions in July 2002 to all access, national long-distance and international long-distance operators to implement carrier selection in their networks (TRAI, August 2008). However, it had not seen the light of day as given by Sridhar (17 October 2005).

Subsequently, the DoT finalized the National Numbering Plan 2003, and mandated that for Subscriber Trunk Dialling (STD, (also referred to as an NLD call) call, the subscriber would have to prefix '010' for selection of the NLD carrier, followed by the Carrier Access Code (CAC), the destination licence area code and the subscriber number (DoT, 2003). Initially, it was proposed that CAC was to be a two-digit code, which would be sufficient for allotment to forty NLD operators. However, the length of the CAC may be reviewed and changed to a three-digit code as and when required.

Despite all these measures, due to the government incumbents' delay, carrier selection was not implemented until 2008. The TRAI released another consultation paper in May 2008 with the objective of unbundling the long-distance service from local access service. Sridhar (28 June 2006) gives details of how secondary long-distance carrier access using calling cards can be potentially very much useful to subscribers in a competitive long-distance market. After the consultation process, the TRAI (August 2008) recommended that carrier selection was difficult to implement due to the incumbent's technology capabilities and costs involved. Hence it was proposed to introduce the provisioning of calling cards by NLD operators directly to end-subscribers. Modification on the above-mentioned clause in the NLD licence was also recommended to allow the NLD operators to sell calling cards. This move is expected to provide some means for NLDOs to directly access subscribers without going through the access service providers. The necessity of implementing the CAC for a healthy and competitive NLD market is explained in Sridhar (11 October 2005).

Call Origination Patterns in Long-distance Network

As indicated earlier in Figure 5.1, NLD calls originate either in a wire-line network or mobile network. Table 5.6 shows the pattern of outgoing calls from mobile access networks which have a subscriber base more than four times that of fixed-line networks.

The four metro areas contribute to a major portion of the NLD outgoing calls. Following are the possible explanations:

1. The need for making NLD calls for subscribers in non-metro areas are less, due to (i) less migration from the areas and (ii) relatives and friends being in the same locations.
2. The perceived higher cost of NLD calls.

Though NLD call rates have come down drastically in recent years, due to the legacy mentioned earlier, NLD calls are seen to be expensive compared to local calls. The NLD jurisdiction is due to the division of the country into different telecom circles due to the political economic decisions in 1994 (see Chapter 2 for details). However, due to increase in competition, both in NLD and local access services, there is a proposal by the DoT to merge the circles into fewer geographical regions. This will blur the distinction between local access and long-distance services in the country.

Table 5.6 Proportion of Mobile Outgoing Calls on the National Long-distance Network

Licence Service Area	Percentage of Outgoing Calls
Category A	10.27
Category B	10.34
Category C	11.58
Metros	23.19

Source: TRAI, October 2008.

ON-NET VERSUS OFF-NET CALLS

As most of the major access providers such as Bharti Airtel, Tata, and Reliance deployed their own NLD network, they started promoting on-net calls much like the famous 'MCI Friends and Family' plan of the US. In 2002, Reliance Communications started their much controversial Dhirubhai Ambani Pioneer offer in which the NLD calls from Reliance to Reliance were marketed at Re 0.40/min when in fact the carrier charge as per the IUC regime was as high as Rs 1.10 (Varghese, 2006). The regulator considered this as predatory pricing and refused to allow this plan of Reliance to be implemented. Table 5.7 gives an overview of the share of on-net/off-net NLD calls:

The share of off-net GSM calls is more compared to CDMA, as there are six to seven GSM operators in every LSA and not all of them have their own NLD infrastructure yet to offer on-net call facilities. On the other hand, the major operators in the CDMA access service (for example, Reliance, BSNL, Tata Indicom) could route the NLD calls through their own network taking economic advantage of the on-net placement of calls. However, with almost all the operators acquiring the NLD licence recently, this situation might change in the future in favour of on-net calls. From Table 5.7 it is clear that the percentage of calls terminating on fixed line networks are minimal, thanks to the rapid growth of mobile services.

Because of the economies of scale effect and shared infrastructure cost, on-net calls will be less expensive for the operators and hence the reduction in cost should be passed on to the subscribers. However, the regulator should keep a watch on whether this will cause a 'tipping point'

Table 5.7   Comparison of Different Types of National Long-distance Calls

Licence Service Area	From GSM Mobile			From CDMA Mobile		
	To Fixed	To Mobile		To Fixed	To Mobile	
		On-net	Off-net		On-net	Off-net
Category A	10%	38%	52%	11%	49%	40%
Category B	9%	29%	62%	8%	51%	41%
Category C	9%	31%	60%	7%	50%	43%
Metros	10%	22%	68%	8%	37%	55%
Total	10%	30%	61%	9%	44%	47%

Source: TRAI, August 2008.

in the market allowing integrated operators (those with access and long distance) to garner substantial market share at the expense of refusing interconnection to niche non-integrated access/long-distance providers.

Comparison of India's National Long-distance Market with that of the United States

At this point, it is important to note the following differences in the NLD market vis-à-vis the US. In the US, the long-distance market was competitive since the 1960s, much before competition was introduced in the local access market (basic telecom service) in 1996. Hence various long-distance companies were at the mercy of the monopoly bottleneck facilities of the local access operators to terminate their calls. Carrier selection (both primary and secondary) was present for the long-distance service and hence provided an opportunity for the long-distance service provider to reach the end-subscribers.

However, in India, the NLD market was opened up for competition much after competitive mobile access services were introduced and even after the new private operators started providing competitive wire-line services. Without carrier selection, the NLD operators could not directly reach subscribers and bill for the service. Hence the need for the IUC levied by the access providers and passed on to the NLD operators, unlike in the US where the long-distance operator passed on access charges to access providers.

International Long-distance Service

The ILD service is defined as a network carriage (also called bearer) service, from India to other countries, through international connectivity provided by different carriers (DoT, 2002). Full flexibility is allowed to the ILD service provider with regard to the type of bearer services offered and transport protocols deployed. The ILD service provider is permitted to offer all types of bearer services from an integrated platform. The ILD service providers can provide bearer services so that end-to-end teleservices such as voice, data, fax, video and multimedia, and so on, can be provided by the access providers to the customers.

The ILD service was provided by the monopoly government operator Videsh Sanchar Nigam Limited (VSNL) until March 2002. On 1 April 2002, VSNL's monopoly was ended prematurely (ahead of the proposed date in 2004) with the Tata group picking up 25 per cent stake in VSNL. Until then VSNL was one of the elite public sector units which was making profits. In 2001-2, VSNL made a profit of about Rs 1,407 crores and paid a record interim dividend of 750 per cent (*Voice and Data*, 2002). Until the divestment, VSNL earned close to 87 per cent of the revenue from ILD voice traffic. The lucrative business of VSNL is described in Sridhar (17 April 2001).

International Long-distance Licence Process

Much like the NLD licenses, the ILD licenses were issued at a fixed entry fee of Rs 25 crores in 2002 for a period of twenty years. Three private operators, namely Bharti Telesonic, Data Access and Reliance Infocomm got the licence to provide ILD services. There were rollout obligations for the ILD licence holders to set up PoP to provide interconnectivity to international networks. Within three years, the licensee had to establish a minimum of four PoPs, one in each region of the country: Eastern, Western, Northern and Southern India. The interconnect facilities should preferably be of open architecture such that interconnections to various NLD operators in the country could be established with the prescribed quality of service norms recommended by the regulator. There was also rollout obligation on connectivity to different parts of the world. Delivery of traffic to all the countries in the world was to be ensured through at least four direct routes, that is, one each to North America, Gulf Region, Europe and any one location in South East Asia, Far East, and Oceania. It should be ensured that traffic to the remaining countries is transited through one of these hubs abroad. It should be possible to terminate traffic to any global destination. The licensees had to deposit Rs 25 crores in the form of bank guarantee to the government so that in case the rollout obligations were not met, it could be encashed. In addition to the entry fee, the licensees were required to pay (including USL) 16 per cent of the AGR to the government in the form of annual licence fee.

In a marked departure from earlier guidelines, for the first time, the ILD licence guidelines acknowledged the emergence of packet-switched technologies for transport of voice. The ILD service providers were permitted to deploy circuit-switched or managed packet-switched network to engineer their ILD networks. However, ILD service providers were also permitted to engineer lower than toll quality network for the customers who did not mind some degradation in the voice quality. The toll quality would mean a Mean Opinion Score (MOS) of 4 or above in a scale of 1–5. This paved the way for some of the ILD licensees (data access in particular) to deploy managed packet-switched networks to carry the voice calls out of/into the country.

Subsequently, in 2005, similar to the NLD, the ILD licence fee was also reduced to Rs 2.5 crores and the annual licence fee reduced to 6 per cent of the AGR (DoT, 2005). With this reduction, many firms acquired the ILD licence and the list of ILD licence holders is given in Table 5.8. The rollout obligations were also lessened which stipulated at least only one gateway switch having appropriate interconnections with at least one NLD operator/access service provider for the receipt and delivery of traffic from/to all the exchanges of the country to be deployed within three years of acquiring the licence.

The International Long-distance Market

The growth of the ILD market is shown in Table 5.9. The ILD market went through a negative growth soon after the entry of private operators. This was due to aggressive pricing adopted by the new entrants. The policy of allowing near-quality ILD service using packet-switching technology encouraged new operators such as Data Access to adopt Voice over Internet Protocol (VoIP) technology for their ILD service. The new ILD operator showed the power of new technology and grabbed about 10.7 per cent share in the very first year of operation (*Voice and Data*, 2003).

However, during this time, the private operators such as Bharti Telesonic, Reliance, and VSNL were deploying ILD infrastructure to carry the calls out of/into the country. Reliance acquired the bankrupt global undersea telecom operator Fibre Link Around the Globe (FLAG) for a mere US$ 211 million. Videsh Sanchar Nigam Limited acquired Tyco Global Network for US$ 136 million and Bharti bought an 8 per cent equity stake in the US$ 500 million, 20,000 km next-generation undersea cable system SEA-ME-WE-4 project (Sridhar, 12 September 2006). Bharti also constructed a 650-mile-long submarine optic fibre cable connecting Chennai, India with Singapore in collaboration with the Singapore-based Network i2i deploying Dense Wave Division Multiplexing technology capable of transport of up to 88 Terabits/sec. Once the infrastructure was built the new entrants were able to bring down the prices which in turn increased the minutes of usage, thus ringing in more revenue for the industry. Even in March 2010, VSNL (now rechristened as Tata Communications) still held more than 50 per cent of the ILD market followed by Bharti and Reliance. Data Access which started off very well, could not keep pace with the infrastructure development and aggressive pricing by the competitors and became bankrupt.

It is to be noted that ILD is the third largest revenue-earning segment, next only to cellular mobile and fixed wire-line services. With the reduction of the entry and annual licence fees, many international operators including British Telecom, AT&T, and Cable & Wireless have started ILD operations in the country. The market share of the different operators is shown in Table 5.10. The effect of increased competition in the ILD segment is presented in Sridhar (13 September 2002).

Table 5.8   List of International Long-distance Licence Holders

S. No.	Name of the Licensees	Licence Signed on
1.	M/s Reliance Communications Limited	25 February 2002
2.	M/s Bharti Airtel Limited	14 March 2002
3.	M/s Data Access (India) Ltd. (Licence under suspension)	27 March 2002
4.	M/s Bharat Sanchar Nigam Ltd	29 January 2003
5.	M/s Videsh Sanchar Nigam Ltd. (Effective from 01.04.02)	5 February 2004
6.	M/s i2i Enterprises Ltd. (BT Global Communications India Pvt. Ltd.)	11 July 2006
7.	M/s AT&T Global Network Services India Pvt. Ltd.	9 October 2006
8.	M/s Vodafone Essar South Ltd.	13 November 2006
9.	M/s Sify Communications Ltd.	21 November 2006
10.	M/s Dishnet Wireless ltd.	13 December 2006
11.	M/s BT Telecom India Pvt. Ltd.	20 February 2007
12.	M/s Tulip IT Services Ltd.	6 July 2007
13.	M/s Spice Communications Ltd.	8 August 2007
14.	M/s Verizon Communications India Private Ltd.	3 January 2008
15.	M/s Cable & Wireless Networks India Private Ltd.	15 February 2008
16.	M/s P3 Technologies Private Limited	28 February 2008
17.	M/s Mahanagar Telephone Nigam Limited	18 June 2008
18.	M/s Equant Network Services India Private Ltd.	20 June 2008
19.	M/s Swan Connect Communications Private Ltd.	12 August 2008
20.	M/s Citicom Networks Private Limited	3 October 2008
21.	M/s Swan Telecom Private Limited (M/s Etisalat DB Telecom Private Limited)	6 October 2008
22.	M/s SingTel Global (India) Private Limited	5 March 2009
23.	M/s Datacom Solutions Private Limited	18 March 2009
24.	M/s Unitech Long Distance Communication Services Limited	28 April 2009

Source: DoT, July 2009.

Table 5.9   Revenue of the International Long-distance Market

Year Ending	Revenue in Rs Million	Y-Y Growth
March 2001	79,660	
March 2002	68,540	
March 2003	54,450	−20.56%
March 2004	43,460	−20.18%
March 2005	38,300	−11.87%
March 2006	72,510	89.32%
March 2007	115,060	58.68%
March 2008	115,320	0.23%
March 2009	150,000	30.07%
March 2010	176,000	17.33%

Source: Voice and Data, July 2010b.

Table 5.10   Market Share of International Long-distance Operators

International Long-distance Operators	Revenue (in Million of Rs)		Y-Y Growth	Market Share
	Ending March 2009	Ending March 2010		
Tata Communications	81,930	92,350	12.72%	52%
Bharti Airtel	26,320	26,000	−1.22%	15%
Reliance	17,290	18,000	4.11%	10%
BSNL	10,520	10,900	3.61%	6%
Vodafone	5,720	8,000	39.86%	5%
British Telecom	2,800	4,540	62.14%	3%
AT&T	3,700	3,980	7.57%	2%
Others	2,530	1,440	−43.08%	1%
Total	150,000	176,000	17.33%	

Source: Voice and Data, July 2010b.

Unlike BSNL which lost a substantial portion of the market share to the new entrants, Tata Communications, the erstwhile government monopoly operator (VSNL) was able to maintain the market share and lead the ILD segment in revenue. This is partly due to the revamping of the operations and human resources by the Tatas, thus transforming the government operator into an efficient and competitive private operator.

International Settlement

In practice, the revenue generated from a call is collected by the telephone company in the country in which the call is originated. This revenue can differ across countries if the demand for international calls is different across these countries. The difference in demand and the actual calls made may vary widely if one of the countries is developed and the other is a developing country. For example, in 1998, India generated 436.2 million call minutes compared to the 1,498.8 million call minutes it received from outside. Moreover, the cost of completing an international voice call includes the cost of originating it by the carrier in the originating country and terminating it by the carrier in the destination country. The costs of origination and termination may be different in different countries, especially so between a developed and a developing country. This suggests that carriers ought to have some way of compensating each other in case there are any imbalances of calls and costs between them. The method of compensating payments between the carriers in two countries is generally a negotiated fixed rate per minute of call duration, normally referred to as the 'international accounting rate'. For example, in 1998, the accounting rate between India and the US was $ 0.62. International accounting rates are the most important components of the marginal cost of international telephone service.

These types of arrangements can be traced back to 1865, when twenty European nations formed a union, now known as the International Telecommunications Union (ITU) (Shy, 2001). Using the accounting rate, the 'net settlement' between carriers in the two countries are worked as follows:

Let N and S be the two countries generating η_N and η_S call minutes per year respectively where $\eta_N >> \eta_S$. If α is the accounting rate settled by the carriers in N and S, then the net settlement paid by the carrier in N to the carrier in S can be calculated as follows:

$$Net\ Settlement_{N \to S} = \alpha (\eta_N - \eta_S) > 0$$

Since the US generates more traffic than most of the countries it pays huge settlement to other countries. For example, in 1996, US carriers paid out in the order of $ 5.5 billion more in such settlements than they received. India thus received $ 431 million in 1998 and received about $ 2.8 billion between 1985 and 2000 from the US alone as net settlement.

This settlement was supposed to be used by the receiving country to improve the telecom infrastructure so that it could slowly bring down the cost of a call, thus improving the outbound call volume and hence the accounting rate.

International Settlement Policy of the United States

In the US, the FCC formally adopted its *International Settlements Policy* (*ISP*) into its rules in the 1980s (FCC, 2008). The *ISP* was initially developed to prevent anticompetitive behaviour on US international routes at a time when, in most countries, telephone service was provided by a monopoly operator. The FCC established the policy to create a unified bargaining position for US carriers because foreign carriers with monopoly power could take advantage of the presence of multiple US carriers by 'whipsawing' or engaging in anticompetitive behaviour. 'Whipsawing' generally involves the abuse of market power by a foreign carrier or a combination of carriers within a foreign market that is intended to play US carriers against one another in order to gain unduly favourable terms and benefits in arrangements for exchange of traffic (FCC, 2008).

While the *ISP* protected US customers from the abuses of market power such as 'whipsawing', international calling rates remained high, in spite of the fact that technological advances and competition were causing US domestic rates to fall. These rates remained high because in many countries, competition was nonexistent or insufficient to drive settlement rates down to cost-based levels. In an effort to drive settlement rates closer to cost, the Commission exercised its jurisdiction over US carriers in 1997 and prohibited them from paying inappropriately high rates to foreign companies to the detriment of US consumers. Specifically, the Commission established its benchmarks policy with the goal of reducing above-cost settlement rates paid by US carriers to foreign carriers for the termination of international traffic, where market forces had not led to that result. The benchmarks policy requires US carriers to negotiate settlement rates at or below benchmark levels set by the Commission in its 1997 Benchmarks Order. The Benchmarks Order divided countries into four groups based upon economic development levels as determined by the information from the ITU and World Bank. As such, the following benchmark rates apply:

1. Upper Income–15¢
2. Upper Middle Income–19¢

3. Lower Middle Income–19¢
4. Lower Income–23¢

As per the above, the benchmark rate for India was set at 23¢. The ITU intervened and proposed a scheme of increasing it to US$0.35 by 2001, thus providing some relief for countries with low teledensity and no competition in the ILD market such as India. Table 5.11 illustrates the international accounting rates and settlement amounts between India and the US over the years. As can be seen from the table, even though the accounting rates were coming down, even by 2002, the rate between India and the US did not reach the benchmark level. Sridhar (8 September 2000) gives details of the effect of lowering international accounting rates on ILD call rates.

However, after the entry of private operators in the ILD market, the accounting rate continued to decline and met the benchmark levels. Subsequently, in its 2004 *ISP Reform Order*, the FCC reformed its rules

Table 5.11 International Settlement Rates between India and the United States

Year	Accounting Rate (in US$)	Net Settlement (in US$)
1985		15,682,263.00
1986		14,911,582.00
1987		7,884,781.00
1988		8,089,022.00
1989		13,799,001.00
1990		20,002,122.00
1991		28,152,949.00
1992		39,776,870.00
1993	0.974	75,337,335.00
1994	0.903	126,194,972.00
1995	0.897	209,707,883.00
1996	0.819	305,290,951.00
1997	0.774	405,690,882.00
1998	0.620	430,842,354.00
1999	0.584	526,326,800.00
2000	0.473	546,441,149.00
2001	0.394	
2002	0.266	

Source: FCC, 2002.

to remove the *ISP* from US international routes for which US carriers have negotiated benchmark-compliant rates. The FCC hoped that lifting the *ISP* on those routes would allow US carriers greater flexibility in negotiating arrangements with foreign carriers. A carrier that sought to add a route to the list of routes exempt from the *ISP* could do so by filing an effective accounting rate modification showing that a US carrier had entered into a benchmark-compliant settlement rate agreement with a foreign carrier that possessed market power in the country at the foreign end of the US-international route that is the subject of the request. India finally entered the list of countries that are exempt from the *ISP* (FCC, 2008).

However, Shy (2001) contends that when international carriers are competitive, an increase in the international accounting rate increases the profit of each company. This is because, in a competitive market, both countries' ILD carriers charge price equal to their marginal costs. Since the only cost of placing an international call is the international settlement fee, competitive pricing means that both carriers set the price of an international call equal to the accounting rate. Under competitive pricing, the profit function of each firm is given by:

$$Profit_N = \alpha \eta_S; Profit_S = \alpha \eta_N$$

Hence the sole source of profit of each ILD carrier comes from access charges. If the regulatory intervention is not present, the above situation might lead to continuously increasing accounting rates. Hence FCC modified the *ISP Reform Order* and clarified safeguards to prevent anticompetitive conduct by carriers. The Commission maintained the 'No Special Concessions Rule' that prohibits US carriers from accepting special concessions from foreign carriers with market power, and established processes for bringing allegations of anticompetitive harm before the FCC. The FCC also said that it would regard certain actions as indications of potential anticompetitive conduct by foreign carriers, including, but not limited to: (i) increasing settlement rates above benchmarks; (ii) establishing rate floors that are above previously negotiated rates, even if below benchmarks; or (iii) threatening or carrying out circuit disruptions in order to achieve rate increases or changes to the terms and conditions of termination agreements. Hopefully, these safeguards will prevent the now competitive ILD markets in most countries to increase the accounting rates well above cost.

Access Charges and Their Ill-effects

The ILD market is subjected to the following charges:

1. Carriage charges specified as ceiling in the IUC regulation.
2. Access deficit charge as explained in Chapter 2, for compensating the basic wire-line service providers for providing high-cost services to low-revenue-earning areas of the country. The ADC on ILD calls was set at Rs 4.25 in 2003 and was periodically reduced to Rs 3.25 in 2005; Rs 1.60 in February 2006; and subsequently to Re 0.50/minute in 2008 (TRAI, March 2008).
3. International accounting rate as explained in the previous section.

In this section we take a close look at ADC which imposes the largest burden on the ILD operator. It is interesting to note that ADCs, as given in (2), have been continuously declining over the years.

The ADC for incoming international calls to a mobile phone was set at more than ten times that of a local call charge at the beginning of the ADC regime in 2003. This high differential forced some ILD operators to bypass the ADC. The ILD operator would masquerade an international call as a local call by removing the international caller identification and substituting it with random digits appended to the local exchange identification tag. By doing this, the ILD operator was able to pass on the ILD incoming call as a local call to the terminating access provider, thus saving the huge difference in ADC between the international and local calls.

What are the implications of ADC from a public finance point of view? First is the substitution effect. The ADC is analogous to a commodity-specific tax, being charge on a specific telecom service (international long distance). Assuming that the demand for international phone calls is elastic and assuming at first, that companies are honest and pay the charge, the imposition of ADC will make such calls more expensive for consumers. In this case, incentives exist for the consumer to substitute international calls with cheaper alternatives such as electronic mail, chat, or Internet telephony call services, available for example, from ISPs, which also bypass the ADC regulatory regime. In all such cases, where possibilities of substitution exist, public finance shows that the imposition of ADC can cause deadweight loss, since there is a loss to the consumer, who substitutes the relatively cheaper, lesser quality service such as Internet telephony for international calls. Since, in this case, the ADC-levied service is not being used, the revenue earned by the government operator is also reduced.

The second effect of the ADC is the evasion of these charges by the operators through masquerading. There could be several reasons for companies to be able to offer low prices on international calls, of which non-payment of ADC is just one sly alternative. There is no doubt that the consumer making international calls stands to save. But the government operator stands to lose some part of revenue from ADC, when masquerading exists.

Now it can be shown and extended from the standard public finance theory that if the ADC has the effect of creating a deadweight loss, its non-payment will add to this cascading effect. This is especially so the case if the revenue loss to the government from non-payment of the ADC was to be higher than the savings the consumer realizes due to a low price for international calls (assuming the low price is attributable to non-payment of the ADC).

Sridhar and Sridhar (1 December 2005) argued, based on standard public finance, that an amount equivalent to the ADC, if imposed as a lump-sum charge (rather than as per minute ADC on specific services such as international calls) on telecom companies, increases revenues. They showed that such a lump-sum charge would cause no consumer substitution that would result from a commodity or service-specific charge. The intuition behind this result is that with a lump-sum charge on a larger revenue base of the companies, higher revenue can be generated, than with a service-specific charge, while allowing consumers to use their preferred service (not substitute cheaper alternatives).

Subsequently, as pointed out in Sridhar and Sridhar (1 December 2005) the TRAI migrated ADC partly, from per minute basis to percentage of revenue of operators. However, the ADC on ILD incoming calls was still kept on a per minute basis, however, it was decreased substantially over the years. As of April 2008, the ADCs on all services except ILD incoming calls were removed. The ADC on incoming calls was kept at a mere Re 0.50/ minute and was slated to be abolished and hence merged with the USO Fund. Box 5.1 illustrates the ill-effects of ADC on ILD calls with specific references to some private operators masquerading ILD calls in 2003–4 (Sridhar, 14 October 2004).

Submarine Cable Landing Rights

Private and public communication of voice, data, image or video across international boundaries are carried through international gateways set

Box 5.1 International Long-distance Turning Grey!

Bharat Sanchar Nigam Limited was in dispute with Reliance Infocomm regarding tampering of the country codes of the incoming ILD calls to make them appear as if they were local calls. There were earlier reports that even the other players such as Bharti and Hutch were also involved in a similar case. The main reason for this kind of masquerading was to save on the ADC of Rs 4.25 per minute levied on international calls to be paid to basic service operators such as BSNL and MTNL. It is estimated that illegal international calls amount to about 2 billion call minutes a year in India, amounting to more than Rs 4,000 crores.

Until recently, only small operators were involved in grey market activities by setting up illegal exchanges. The calls used to be routed through the private/satellite network or through the Internet, bypassing the ILD gateways. This caused loss of revenue and hence was a concern for legal operators (read ILD licence holders). For the government it meant loss of licence fee which is 15 per cent of the adjusted gross revenue of ILD services. However, the involvement of ILD licence holders themselves (for example, Reliance) in smuggling the calls should be strongly reprimanded.

The TRAI, to whom the case was referred to by BSNL, said that it had no vigilance power or investigative infrastructure to check the grey traffic and the matter was forwarded to the DoT. The TRAI should have recommended a policy (much like what the Sri Lankan regulator did) making it mandatory for all ILD operators to provide details of all traffic handled and all interconnection agreements periodically for monitoring purposes. The TRAI should also be empowered to make policies (instead of DoT making policies currently!) for curbing grey traffic and enforcing penalties for smuggling. The TDSAT should be referred for addressing concerns of the basic service operators.

Or, is the ADC itself to be blamed for these illegal activities? The main purpose of the ADC was to cross-subsidize the price of the basic services and was arrived at by the TRAI in its IUC regulation, mainly based on the historical cost data provided by BSNL. Due to changes in technology, the cost of providing basic services has reduced, thus reducing the need for such cross-subsidization. The USO Fund has grown to cover the cost of providing services in rural areas.

up by respective operators, either through submarine cables or through the satellite system. In this section, we look at communication through submarine cable systems and the issues therein. International Private Leased Circuit (IPLC) is one of the significant elements of international connectivity for the Internet, and secure and reliable cross-border communication for firms such as software companies and information technology (IT)-enabled services. Such international connectivity can be provided through submarine cable systems from the coastal points of the countries. An international submarine cable system can normally be divided into the 'wet' portion of submarine cables, the landing stations or 'headends', and backhaul facilities for domestic connectivity.

The 'wet' portion corresponds to the submarine optic fibre cable itself. The construction, provisioning and maintenance support of the cable facilities involve long lead times and high sunk costs. The process of planning and installing a cable system is very complicated and includes getting approvals from various national agencies. The cable landing stations are points at which the international submarine cables come onshore and terminate. Generally, these are buildings, which contain the onshore end of the submarine fibre optic cable, house the necessary equipment to interconnect and pass traffic to and from the submarine cable, and are the point where the submarine cable capacity is connected to the domestic backhaul circuit. The backhaul facility is the high-capacity inland domestic circuit, which is required by service providers to link the cable landing station to their existing national infrastructure. This is very similar to the Domestic Leased Circuits (DLCs) used by the NLDOs. A simplified schematic diagram of the submarine cable system is shown in Figure 5.3.

The historic approach to the creation of an undersea cable was to form a closed club of operators that would raise the capital for the investment needed to lay the cable. The members then would have the exclusive rights to use that capacity in their respective countries (Esselaar *et al.*, 2007). An example of such a recently constructed cable system is the SEA-ME-WE 4 (South East Asia–Middle East–Western Europe) cable

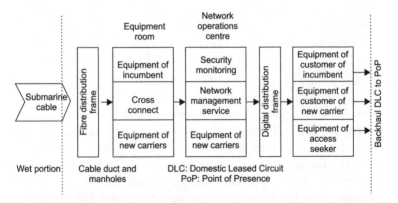

Figure 5.3   Schematic Diagram of Submarine Cable System

system developed by a consortium of sixteen telecom companies. After construction, other operators would be allowed to purchase an Indefeasible Right of Use (IRU), a contract transferring to them the right to use a dedicated amount of capacity on the undersea cable system. While the non-club IRU owner does not bear the risks of the construction process, the unit price for the capacity it pays is usually higher than for the club members. Table 5.12 lists the different submarine cables terminating in India and their associated cable landing stations and capacities.

With a total installed capacity of 18.6 Tbps, the international bandwidth landing at Indian shores is huge compared to the list capacity of 588 Gbps (a mere 3 per cent of the design capacity). The architecture of submarine cable systems such as SEA-ME-WE 4 can be accessed on the Internet.

It is cited by many policymaking bodies that one of the underlying causes of the high costs of telecom in developing countries is due to high charges for international capacity between countries, especially to other continents (Esselaar *et al.*, 2007). After the ILD market was opened up for competition, access to the bottleneck cable landing facilities owned by the erstwhile monopoly operator became an important issue. Box 5.2 provides an example of such a case.

In order to provide non-discriminatory access to bottleneck submarine cable landing facilities and to reduce cost to end-users, the TRAI initiated process in 2004, for regulating the price of IPLCs. Videsh Sanchar Nigam Limited strongly resisted the regulation process and filed a case against the TRAI in the TDSAT. However, after a number of dispute resolution processes, the TRAI was able to fix the price for IPLC half-circuits using a cost-based tariff for an E1 (2 Mbps) at Rs 1,300,000 per annum and used ratios of 1:8:23 for E1, DS3 and STM1 lines (TRAI, September 2005). Though discounts can be given on the tariff, the TRAI mandated them to be transparent and non-discriminatory. This brought down the costs of IPLCs by as much as 59 per cent for higher capacities.

The physical interconnection is necessary for provision of access to the international submarine cables. The new entrants will require access to essential support facilities and unbundled network elements at the following points of access:

1. Cable Duct and Manholes
2. Fibre Distribution Frame
3. Equipment Room
4. Network Operations Centre

Table 5.12 Capacity of Submarine Cables in India

S. No.	Submarine Cable	Landing Station	Owner of Landing Station	Type of Cable System	Designed Capacity (Gbps)	Equipped Capacity (Gbps)	Utilized Capacity (Gbps)	Spare Capacity (Gbps)
1.	SMW3	Mumbai	VSNL	Consortium, Protected	212	20	20	0
2.	SMW4	Chennai, Mumbai	Bharti + VSNL	Consortium, Protected	1,200	40	27	13
3.	SAFE	Cochin	VSNL	Consortium, Unprotected	5	5	5	0
4.	FLAG (Reliance)	Mumbai	VSNL	Hybrid, Protected	160	20	10	10
5.	i2i	Chennai, Mumbai	Bharti + VSNL	Private, Unprotected	8,400	160	12	148
6.	TIC	Chennai, Mumbai	VSNL	Private, Unprotected	5,100	320	16	304
7.	Falcon	Mumbai	Reliance	Private, Unprotected	2,560	2.5	1	1.5
8.	Indo–Sri Lanka Cable	Tuticorin	BSNL	Private, Unprotected	960	20	–	20
	Total				18,597	588	91	497

Source: TRAI, May 2008.

Box 5.2 Videsh Sanchar Nigam Limited–Reliance Dispute over Landing Rights of Fibre Link Around the Globe

The erstwhile monopoly ILD operator VSNL had exclusive landing rights from the submarine cable operator FLAG which terminated its undersea cable at VSNL's landing station at Mumbai, India. Meanwhile, rival ILD operator Reliance acquired FLAG in January 2004 for about US$ 211 million. In May 2004, the two companies agreed to share the existing capacity of the cable landing station (VSNL, 2005). However, thereafter VSNL denied FLAG permission to upgrade the capacity from its existing 10 Gbps. Reliance (also referred to as the Anil Dhirubhai Ambani Group (ADAG)) filed for arbitration with the international tribunal in December 2004 against VSNL (also referred to as Tata Communications) on two accounts. One, it had asked for directions to allow the company to upgrade the capacity of the Indian leg of the cable. Two, it demanded compensation for the business opportunity it had lost due to its inability to upgrade the Reliance capacity (*Business Standard*, 2008). A Netherlands-based district court in the Hague upheld an order by the arbitration tribunal of the International Chamber of Commerce in 2006 directing VSNL to permit Reliance Globalcom (formerly FLAG Telecom) to upgrade its bandwidth capacity at the cable landing station in Mumbai. The district court also ordered VSNL to pay proceeding charges of Euro 13,092 plus Euro 12,844 for legal representation to ADAG.

5. Digital Distribution Frame
6. Backhaul Transmission
7. Landing Facilities

Moreover, the new entrants might require co-location facilities that include building space and power. Site maintenance and environment services are also required at the cable landing stations. Since the cable landing station is a bottleneck facility, the dominant/incumbent operator has to take reasonable measures to accommodate new operators. The cost of access and co-location space and associated expenditure must be recovered in an equitable manner from various operators.

Realizing this, the TRAI recommended mandatory sharing of cable landing stations (CLSs) in its recommendations (TRAI, December 2005), which was further accepted by the DoT. Further, the government amended ILD licenses in 2006 for enabling the TRAI to bring out regulations to ensure efficient, transparent and non-discriminatory access to CLSs. Following this, the TRAI launched a further consultation in December 2006 on measures concerning the resale of IPLCs. While it acknowledged that there were economic limits on the number of facilities-based operators, it noted the enormous differences in the number of resellers in other countries. Hence to further increase competition in the IPLC segment, the TRAI released its recommendations in March 2007, which were subsequently accepted by the Telecom Commission, allowing resale of international bandwidth to resellers. The scope of resellers is as per the following (TRAI, March 2007):

1. Provide end-to-end IPLC between India and the country of destination for any capacity denomination.
2. Public switched telephone network /public network not to be connected with IPLC.
3. The resellers to take IPLC from the licensed ILD service providers (ILDOs).
4. Resellers shall be permitted to enter into an arrangement for leased line with access providers, NLD service providers and ILD service providers for provision of IPLC to end-consumers.
5. The resellers can access the subscribers for provision of IPLC only and not for any other purpose.
6. Resellers should be allowed to sell bandwidth on a retail basis with or without value addition to end-consumers. Resellers should not sell the bandwidth to other resellers.
7. Co-location of the equipment of the resellers at CLS should be on the same terms and conditions as may be made applicable from time to time for licensed ILD operators.
8. Resellers should be allowed to create their own infrastructure for multiplexing, de-multiplexing, billing, and customer management.
9. Licence of the resellers should be for the whole country and not limited to any single service area.

No ceiling was prescribed for resellers of IPLCs. An entry fee of Rs 1 crore with an annual revenue sharing of 6 per cent of AGR was levied on the resellers. Allowing resale would encourage licensed ILDOs to resell the excess bandwidth on their international circuits to newcomers such as foreign companies who provide enterprise network services. Though Bharti announced that it was willing to resell, the effect of the above regulatory intervention is yet to see the light of day even almost two years after it was announced. This further requires that collocation and unbundling of the CLS is open for resellers as well. Finally in June 2007, the TRAI issued the International Telecommunication Access to

Essential Facilities at Cable Landing Station Regulation, the essential features of which are listed below:

1. Every owner of the CLS is mandated to submit Reference Interconnect Order (RIO) for approval of the Authority within thirty days.
2. Owners of the CLSs have also been asked to provide the costing details in arriving at the various charges submitted to the Authority as a part of the schedule of their document.
3. Provision of access to the CLS by the owner of the CLS shall be on non-discriminatory basis.

Apart from the above, the procedures for provisioning and access to reference capacity, and co-location of sites and space requirements, and agreements were also streamlined. With this regulation, any bottleneck on the CLS facility is effectively removed. The summary of the above reforms is given in Table 5.13.

Table 5.13 Regulatory Changes on International Long-distance, Cable Landing Stations, and Access to Submarine Cable Capacity

Regulation/Policy	Date
ILD segment opened up for competition	1 April 2002
Tariff fixing for IPLC	September 2005
Reduction in ILD entry and annual licence fees	December 2005
TRAI recommendation on mandatory sharing of CLSs	December 2005
TRAI recommendation on resale of IPLC	March 2007
TRAI regulation on CLS	June 2007

* * *

The monopoly NLD and ILD services migrated to a competitive market. Though a number of licence holders are providing service, it is time that the policymakers and regulator implement the carrier access code to promote customer choice of carriers.

6

Internet Services
Regulations Stifling the Growth

History of the Internet

The Internet is a network of networks and was started as a project on Advanced Research Project Agency Network (ARPANET) of the Department of Defence (DOD) in the US in 1969. The objective behind the project was to build a reliable network called ARPANET with distributed computing power which could withstand calamities. The ARPANET was the first operational packet-switched network with four operational centres at University of California, Berkeley, University of Santa Barbara, University of Utah, and the Stanford Research Institute in the US. The widely used protocols for different devices to communicate over the Internet, namely Transmission Control Protocol (TCP) and the Internet Protocol (IP) were developed by them in 1974. Box 6.1 gives details on the history of the Internet. Table 6.1 lists the events in Internet services, and their corresponding timelines.

Growth of the Internet

The Internet subscriber base and the hosts connected to the Internet continue to grow, worldwide. Although the US has led the way in the diffusion of the Internet, Internet diffusion has also become a worldwide phenomenon. The growth in number of hosts (or devices) connected to the Internet is shown in Figure 6.2.

Box 6.1   History of the Internet

In 1972, Ray Tomlinson of Bolt Beranck and Newman (BBN) wrote the first system to provide distributed mail service across a computer network using multiple computers (Stallings, 2001). Vint Cerf and Bob Kahn of ARPA started developing methods and protocols to internetwork different computer systems. The University of California, Berkeley incorporated the TCP/IP protocols as part of its Berkeley UNIX operating system in the early 1980s.

At the same time, the National Science Foundation (NSF) realized the benefits ARPANET had on research and decided to build a successor to ARPANET that would be open to university and research groups. The NSF then started supporting networks of different research communities using the NSFNET backbone. Though originally the NSFNET was designed to interconnect the six supercomputing facilities in the US, eventually the NSF offered interconnection through its backbone to regional packet-switched networks across the country. In 1989, the NSFNET backbone was reengineered to high-speed redundant and reliable transmission links operating at T1 (1.544 Mbps) speed. In 1990 ARPANET was shut down. At the same time, a consortium called Advanced Network Services (ANS) formed by companies Merit, MCI, and IBM took care of the operations of the NSFNET. More and more educational institutions got connected to the Internet not only in the US but also in rest of the world. However, much like the NSF subsidizing the Internet (NSFNET) backbone, various national governments subsidized the Internet backbone for their respective countries.

In 1991, the US government announced that it would no longer subsidize the Internet after 1995. As part of the privatization plan, the US government mandated interconnection points called Network Access Points (NAPs). In 1993, the backbone was running at T3 (45 Mbps) speed and the usage of the Internet was growing. In 1996, the next-generation Internet (NGI) initiative was announced to develop advanced networking technologies and applications on test bed networks that were 100 to 1,000 times faster. Internet2 is the latest test bed for the Internet. It is a collaborative project sponsored by the University Corporation for Advanced Internet Development, a consortium of over 180 US universities that are developing advanced Internet technologies and applications to support research and higher education. (For further details on the history of the Internet, refer to ICRA, 2000 and Sheldon, 2001).

The current architecture allows users at home or offices to connect to the Internet through a local ISP. The ISPs are typically connected by 'wholesalers', who are called the 'Network Service Providers'. These wholesalers in turn connect using Internet connection points called 'Internet Exchanges' to the Internet backbone (Stallings, 2001). The ISPs can also have peering arrangement to interconnect their networks. The general architecture of the Internet indicating the above is shown in Figure 6.1.

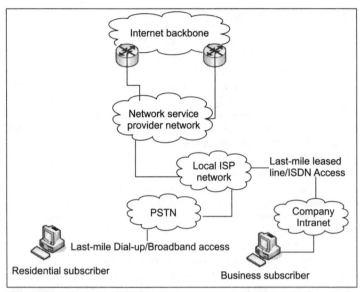

Figure 6.1   Internet Architecture

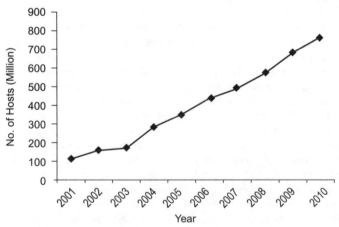

Figure 6.2   Internet Host Count Worldwide

Source: http://www.isc.org

Table 6.1   Events in Internet Services

Year	Event
1969	ARPANET deployed by Department of Defence in the US
1974	TCP/IP Protocol invented at the University of California, Berkeley, US
1995	VSNL launched Internet services in India
1998	Internet Engineering Task Force (IETF) accepted IPv6 draft standard
November 1998	ISP licenses issued to private operators
December 1998	World Intellectual Property Organization (WIPO) domain name dispute resolution centre set up
1 April 2002	Restricted form of Internet telephony opened up in India
2003	National Internet eXchange of India (NIXI) set up
2004	National broadband policy announced
2004	Deployment of any access technologies by ISPs recommended by the TRAI, subsequently implemented by the DoT
April 2004	First recommendations by the TRAI on local loop unbundling
2004	The DoT issued Virtual Private Network (VPN) licence guidelines
2005	.in Domain Name Dispute Resolution policy released
March 2006	Licence fee incorporated for Internet telephony services
March 2006	Unrestricted Internet telephony allowed for UASPs
2007	New ISP licensing policy announced by the DoT
August 2008	The TRAI recommended unrestricted Internet telephony for ISPs, which was subsequently rejected by the DoT
21 December 2010	Net neutrality rules announced by the FCC in the US
2010	The TRAI recommendations on the National Broadband Plan released

Figure 6.3 illustrates how the users of the Internet have grown in the last decade. Internet World Stats (2010) also reports that Asia accounted for 42 per cent of the Internet users compared to Europe's 24.2 per cent and North America's 13.5 per cent. Figures 6.2 and 6.3 indicate that the number of devices connected to the Internet and the users of Internet are growing exponentially. The number of Internet users is increasing substantially in Asia.

The growth of the Internet subscriber base in India since 1995 is shown in Figure 6.4.

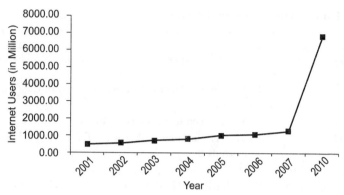

Figure 6.3 Growth of Internet Users Worldwide
Source: www. Internetworldstats.com

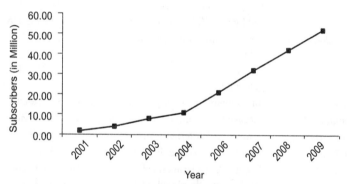

Figure 6.4   Growth of Internet Subscriber Base in India
Source: TRAI, 2010

It must be pointed out that the subscription numbers reflect only the actual numbers subscribing to Internet services and do not reflect the number of users of Internet services. There are a significant number of users who visit cyber cafés and access Internet services on the shared access link.

Most of the ISPs also provided Internet services to businesses using leased line. The number of leased lines providing Internet services to businesses grew from 12,000 in December 2004 to more than 18,000 in December 2006.

Liberalization of Internet Service Provisioning in India

Internet services were first launched in India on 15 August 1995 by the VSNL, the then government monopoly operator. During the first three years of VSNL's operation, the Internet subscriber base grew slowly. By the end of March 1998, it had barely reached 140,000 subscribers (TRAI, May 2007). In order to promote wider availability of Internet services, the Indian government announced liberalization of Internet services thus paving the way for private companies to enter into the provisioning of Internet services. In November 1998, ISP licenses were issued to private operators. The entry conditions for the new operators were very liberal and hence congruent with the philosophy of the Internet. The conditions of licensing are given in Table 6.2.

There was no entry fee, no licence fee, no restriction on number of service providers in a licence area and liberal rollout obligations to bring in as many service providers as possible. The ISPs were permitted to determine their own tariffs. They were also permitted to set up their own international Internet gateways for carrying the Internet traffic in and out of the country (refer to TRAI, May 2007 for details on the licence conditions).

As can be noticed, the policy for Internet services were very liberal compared to the telecom policies existent in 1998. Furthermore, the licensing method curiously coupled the licence areas with the domain operated by telecom companies, especially those providing fixed wire-line basic services. The implicit assumption was that the last mile access from

Table 6.2   Internet Service Licensing Conditions

Category of ISP	Operating Area	Performance Bank Guarantee (in Rs)
A	All India	2 crore
B	Any of the twenty territorial telecom circles (analogous to States) or any of the four metro telephone districts of Delhi, Mumbai, Kolkata, and Chennai and any of the four major telephone districts of Ahmedabad, Bangaluru, Hyderabad, and Pune	20 lakh
C	Any secondary switching area (SSA)* of the DoT	3 lakh

Note: *DoT has about 250 SSAs which come under the twenty-one telecom circles. The list of SSAs is given in http://www.bsnl.co.in/map.htm

the subscriber premise to the PoP of the ISP could be provided only by the access providers such as fixed wire-line basic service operators.

The liberal policy allowed as many as 175 applicants to get ISP licenses. The entry of a large number of players resulted in lower Internet tariffs which led to huge growth in the subscriber base. The Internet subscriber base grew more than 200 per cent from 0.28 million in March 1999 to 3.04 million in March 2001 as illustrated in Figure 6.3. The HHI which is a measure of competition was as low as 0.15 in June 2003 (TRAI, May 2007).

However, the growth of the Internet in the country continued to be sluggish compared to other countries. Though about 700 licenses were issued within the first three years of opening of the ISP sector, only 389 licensees existed in March 2007. Out of these only 139 were functionally active, thus providing service.

The overall Internet services' revenue during 2006–7 was estimated to be Rs 2,040 crores (*Voice and Data*, 2007). This pales in comparison with the revenue of fixed-line services (Rs 30,190 crores), mobile services (Rs 56,183 crores) and even the NLD and ILD services market.

Until 2001, almost all the subscribers connected to the Internet using the dial-up modem service over the telephone network of BTOs. After the launch of the Broadband Policy in 2004, there was growth in the broadband Internet subscriber base in India. Newer technologies such as DSL, cable modem and metro Ethernet were implemented by the ISPs as well as telecom companies (basic service licence and UASL holders) to provide Internet services to residential and business subscribers. Details on various Internet access technologies are provided in Table 6.3.

The BTOs, especially the new entrants, started bundling 'always on' broadband Internet connection with their voice services. The government operators, namely BSNL and MTNL who own more than 90 per cent of the fixed lines in the country started offering broadband Internet connection to their subscribers with aggressive prices. As on June 2007, about 65 per cent of broadband Internet subscribers were with the above two government operators. Of the top ten ISPs, four were access providers (basic or UASL holders); five were pure-play ISPs and one was a cable operator.

Competition reduced the tariff for broadband services to about Rs 250 per month in June 2007 which was one of the lowest in the world. As of September 2007, 47.91 per cent of the Internet subscribers still used the dial-up mode, while 20.09 per cent used DSL broadband connections to access the Internet. This trend was quite different from the international

Table 6.3 Comparison of Last-mile Technologies for Internet Connectivity

Technology	Maximum Bandwidth	Distance Limitation	Characteristics	Service Provider
Integrated services digital network (ISDN)	128 Kbps – 2 Mbps	Up to 5.5 km	Use of existing telephone lines; most of the BTOs in India, especially in cities offer the service; suitable for residential and SMBs	BTO
Digital subscriber loop (DSL)	256 Kbps – 6 Mbps	Up to 5.5 km	Use of existing telephone lines; provided by existing BTOs in most of the cities	BTO
Leased line	256 Kbps – 100 Mbps	Up to 5.5 km (if coppercable); up to 20 km (if optic fibre)	Mainly used by business subscribers; in case of optic fibre: (i) specialized customer premise equipment (CPE) is required and (ii) very high capital expenditure (in the range of US$ 6–10 per meter) for laying down fibre	BTO (if copper/ optic fibre)/ ISP (if optic fibre; ISPs allowed to provide using copper cables in April 2004)
Metro Ethernet	256 Kbps – 1 Gbps	Up to 5.5 km	Use of copper cable or optic fibre	ISP
Cable modem/ Ethernet over cable network	256 Kbps – 30 Mbps	Up to 5.5 km	Deployed by local cable operators using open over-the-air wiring; can use cable modem technology to provide Internet access; copper cables can be hung along the cable network to provide Ethernet connectivity as well	Local cable operator
Terrestrial radio	64 Kbps – 32 Mbps	20-30 km	Transmission radio towers at line-of-sight; specialized CPE required (can be rented or bought); low capital expenditure compared to optic fibre	ISP
VSAT	400 Kbps – 32 Mbps	No distance limitations	Specialized CPE required; high capital expenditure; licence required from the government for operation; long-term service commitment; very high delay (about 500 ms round-trip time)	VSAT service provider/ ISP/ DTH operators

trend where 64 per cent subscribers used DSL and 24 per cent subscribers used cable modem to access the Internet. However, the DSL broadband connections continued to show growth.

Aggressive price competition has also reduced the ARPU of the ISPs. The monthly ARPU which was Rs 400 in March 2004 came down to Rs 200 in March 2005 and continued to decline.

Last-mile Access to the Internet

Subscriber premise is connected to the PoP of the ISP through an access loop normally referred to as the 'last-mile' connection as illustrated in Figure 6.1. The last-mile connectivity is typically provided by the BTO (refer to Chapter 2 for details on basic telecom operations). Initially, the licence allowed ISPs to provide last-mile access in the local area either through optic fibre or radio communication (ICRA, 2000). This was to protect the interests of the BTOs who had paid much higher charges to get their licence compared to the ISPs. Optic fibre was too expensive to be deployed as a last-mile option. Hence most of the ISPs provided access through radio links or had to depend on the BTOs for the last-mile access to the subscriber premise. A look at Table 6.3 indicates that ISPs needed to depend on other access providers, mainly the BTOs for their last-mile access.

The ISPs were also allowed to tie up with authorized Local Cable Operators (LCOs) to provide Internet service. Though some ISPs used this model to provide Internet service, it was limited due to unorganized franchisee model used by the LCOs (for details on LCOs and their market refer to Chapter 8).

However, in order to give impetus to the growth of the Internet in India, the TRAI recommended in 2004 that any access technologies including optic fibre, radio and copper cable be deployed by the ISPs. This was later accepted by the DoT and policies for last-mile access were modified (TRAI, 2007). However, laying down copper cables was expensive for the ISPs. The lion's share of the local loop which connects the customer premise to the BTO's switch through traditional copper wires is owned by government operators. Since the local loops require huge sunk costs, network duplication is neither compatible with the incentives of competitors nor efficient from a social point of view. Hence the ISPs did not deploy the last-mile connection to subscribers and hence the BTOs had an edge over the ISPs in providing bundled Internet services.

Local Loop Unbundling

Instead of the ISP establishing its own local loop or waiting for resources to be made available by the BTO for its exclusive use, it would be more efficient if the ISPs and BTOs collaboratively offer both voice and Internet services through the same local loop. Hence an alternative to this is 'Local Loop Unbundling (LLU)' and is being vigorously promoted as a major regulatory initiative by all countries with a competitive basic services market. The TRAI also recommended LLU for improving competition in broadband service, way back in April 2004 (TRAI, April 2004). Unbundling refers to the process in which the incumbent BTOs lease, wholly or in part, the local loop components of their telecom network to other carriers or service providers. The LLU, in principle, provides new entrants access to the local loop and encourages the provisioning of complementary services such as high-speed Internet connectivity and Internet telephony. Unbundling, as a policy is built on the recognition that incumbent BTOs have a dominant position in the provisioning of local telecom services by virtue of their control over the local loop. Most often it cannot be economically replicated by alternative service providers. Sridhar (2004) explains why LLU will improve the Internet and broadband penetration in India (for details on different forms of LLU, refer to Chapter 2).

However, even by December 2010, the government had not implemented the TRAI's recommendations on LLU. This was mainly due to the resistance by the incumbent government BTOs to give up rights over the use of expensive local loop. Some ISPs have tied up with authorized LCOs as indicated in the ISP guidelines to provide Internet access to retail customers with mixed success. The cable TV industry is fragmented in India and the quality of the network is not ideal for Internet services. Hence the pure-play ISPs started focussing more on business and enterprise customers with their value-added service offerings compared to the plain vanilla Internet service for retail customers.

Internet Telephony

Internet telephony shall be defined as, computer to computer, computer to phone, and phone to phone voice communication, carried over the public Internet, while the traffic carried over a privately managed network using the TCP/IP protocol is referred to as Voice over Internet Protocol (VoIP). The Internet protocol is used for carrying voice over an unmanaged Internet environment. Internet telephony was allowed and prevalent in many countries. However, it was banned in India until it was opened up with some restrictions on 1 April 2002. Existing ISPs were permitted to offer this service after signing the Internet Telephony Service Provider (ITSP) licence. The different ways by which Internet telephony could be provided are illustrated in Box 6.2. Restrictions on Internet telephony services as per the ITSP licence guidelines are given in Table 6.4 (TRAI, 2007).

A PC to PC phone call, which can be carried solely over the public Internet, can be provided by the ISPs and there is no way to monitor and regulate this service. A PC to phone or a phone to phone call can be carried by the ISP on a public Internet but has to necessarily originate/ end on the PSTN or Public Land Mobile Network (PLMN) through the media gateways of the ISPs as illustrated in Figures 2 and 3 in Box 6.2.

Box 6.2   Different Forms of Internet Telephony

Figure 1 describes communication between a PC connected through the ISP access node to the Internet and another PC connected to the Internet in the same way. Users can make calls through the Internet using Internet telephony or Instant Messaging software (most of them freely downloadable). Companies such as Skype, Yahoo! and Vonage have been providing this form of communication services for quite some time. The PCs can be connected to the Internet from any part of the world and hence this form of communication service can never be regulated.

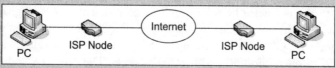

Figure 1   PC to PC Internet Telephony Communication

Figure 2 illustrates how communication can take place between a PC connected to the Internet and a phone connected to PSTN/PLMN. For this form of communication to take place, a media gateway which provides interconnection between the PSTN/PLMN and the Internet is required. This gateway also coverts the voice coming through the traditional circuit-switched PSTN/PLMN to data packets to be transmitted through the Internet. The media gateway also translates the E.164 telephone number of the voice call to the corresponding IP address so that the packets can be routed to the destination PC. Users can download the Internet telephony software and make calls from their PCs to phones using the services of not only Indian ITSPs but also those of foreign service providers such as Skype or Yahoo!

Countries in Europe and the US do not have any restrictions on Internet telephony, and hence media gateways could be set up by the operators to originate and terminate Internet telephony calls. However, as per the restricted Internet telephony regulations, setting up of media gateways in India is not allowed (DoT, 2007). This prevents an Internet telephony call to be terminated on a phone in India. Hence as per the guidelines, an Internet telephony call originating outside India is not allowed to terminate on a landline or mobile phone. Also, any PC to landline/mobile call within India is also not allowed.

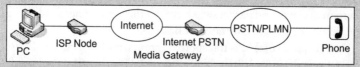

Figure 2   PC to Phone Internet Telephony Communication

In the form of communication illustrated in Figure 3, a normal landline/mobile phone user connected to a PSTN/PLMN can make a call through the Internet to another landline/mobile phone user. This requires the use of media gateways at both the origin and destination of the calls. Since setting up media gateways within India is not allowed, this form of Internet telephony is not allowed.

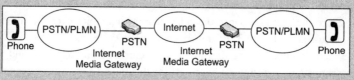

Figure 3   Phone to Phone Internet Telephony Communication

Table 6.4   Restrictions on Internet Telephony Service

Origin	Destination	Allowed	Reason
PC in India/Abroad	PC Abroad/in India	Allowed	
PC in India	Phone Abroad	Allowed	
IP-based H.323/ session initiation protocol (SIP) terminal connected directly to ISP node	IP-based H.323/ SIP terminal connected directly to ISP node within/ outside India	Allowed	
PC in India	Phone in India	Not allowed	NLD network may be bypassed
PC Abroad	Phone in India	Not allowed	ILD network may be bypassed

As per the ISP guidelines, the ITSPs were not allowed to deploy media gateways in India and the reasons are described below (DoT, 2005).

RESTRICTIONS ON INTERNET TELEPHONY

In Internet telephony, real-time voice conversations are converted into data packets and delivered over the Internet which uses packet-switching technologies. This is in contrast to voice calls made through the PSTN/PLMN which uses circuit-switching technologies. The schematic diagram of circuit and packet switching and the corresponding advantages and disadvantages are illustrated in Box 6.3.

As indicated, voice calls transported through packet-switched Internet suffer from delay and jitter. They are of inferior quality compared to the circuit-switched telephony service offered by the BTOs or mobile operators. Hence the TRAI and DoT always indicated that Internet telephony is a different service in its scope and nature from real-time voice as offered by other licensed operators like BTOs, CMSPs, UASLs, NLDOs, and ILDOs (DoT, 2007). This is one of the other reasons for the asymmetric regulation between traditional voice telephony and Internet telephony.

Since the cost and hence the price of providing voice over the Internet as discussed in Box 6.3 are much lower than voice carried over PSTN, the incumbents create hurdles to the new entrants. The regulator and the government, in their interest to protect the incumbent government operators providing basic, long-distance and international long-distance service, also put in the following conditions for the Internet telephony service:

1. The addressing scheme for Internet telephony shall only conform to the IP addressing scheme of the Internet Assigned Numbers Authority (IANA), exclusive of the national numbering scheme/plan applicable to subscribers of the basic/cellular telephony service.
2. The Internet service licensee is not permitted to have PSTN/PLMN connectivity.

The above clauses clearly separated Internet telephony services from the traditional voice service offered through the PSTN/PLMN, through regulatory guidelines. It must be pointed out that such regulatory restrictions are not present in most of the countries including the US and Europe.

If the technology allows (as per the regulation) calls from a PC in Delhi to a phone in New York the same technology allows a call to be

Box 6.3 Circuit and Packet-switching

A network consists of a source from where the information is originated; destination where the information is terminated; intermediate nodes which possibly store, process and forward the information from the source towards the destination; and the transmission links which interconnect the source, destination and nodes.

Circuit-switching

In a network shown in Figure 4, the source A (a phone) is connected to destination B (also a phone) through intermediate exchanges 1–5, deployed in end-offices and intermediate offices of the PSTN/ PLMN. Traditional voice service through the PSTN/PLMN uses circuit-switching for switching the voice calls from A to B. A physical circuit is set up at the time of call initiation by the network which is dedicated for the conversation between A and B. The circuit is released only after the termination of the call.

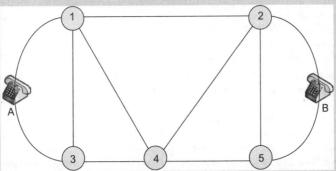

Figure 4 Circuit-switched Network

Packet-switching

As shown in Figure 5, in a packet-switched network, the source divides the information into a large number of discrete units called 'packets'. These packets are then transported from source A (normally a PC) to destination B. The packets are stored and forwarded through the intermediate nodes as and when capacity on outgoing links become available. It is possible that each packet can follow a separate route from A to B (also referred to as *datagram* routing) as is shown in Figure 5 and routes for each packet can be set up dynamically by the network. For example, while Packet 1 follows A-3-4-2-B, Packet 2 follows A-1-2-5-B and Packet 3 is routed along A-1-4-5-B.

Since packet-switching uses store-and-forward philosophy and a route is never dedicated as in circuit-switching, the network efficiency is high in packet-switched networks. This increased efficiency accommodates more

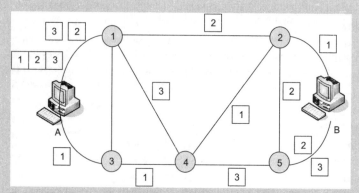

Figure 5 Packet-switched Network

traffic than the circuit-switched networks of the same capacity. Hence the cost of transmitting per unit of information in a packet-switched network is considerably less than in a circuit-switched network. Packet-switched networks can respond to failures and network congestion much better than circuit-switched networks as routing of packets is done dynamically. However, the packets in datagram-routed packet-switched networks can reach the destination out-of-sequence, mechanisms such as addressing and sequencing have to be put in place to put the packets in order at the destination. This also mandates that the end-devices such as A and B are not dumb terminals or phones but relatively expensive intelligent devices (such as PCs or Internet phones) that implement the necessary protocols for processing packets. The store-and-forward and dynamic routing in packet-switched networks results in delay and variation delay of packets reaching the destination from the source to the destination through the network.

Hence while circuit-switched networks are ideal for synchronous communication services such as telephony and video conferencing, packet-switched networks are good for transporting asynchronous data and image communication.

The Internet uses datagram routing-based packet-switching to send packets of information from the source to the destination. Since packet-switching is much more efficient than circuit-switched networks and it can accommodate many more simultaneous connections than circuit-switched networks, it is less costly to send a message through a packet-switched network compared to a circuit-switched network. Moreover, communication over a packet-switched network is less distance-sensitive compared to circuit-switched networks. However, the quality of synchronous communication such as voice over packet-switched networks will be inferior to circuit-switched networks.

made from a PC in New York to a phone in Delhi. Or for that matter it also allows a call from a PC in Delhi to a phone in Kanyakumari, the southern tip of India, for almost the same cost. Even though the technology allows a PC to a phone call at a much cheaper price, thanks to the regulation the subscribers are deprived of it. The above restrictions on Internet telephony disallowing connectivity between the Internet and the PSTN are due to the lobbying by the NLD and ILD operators who feared calls bypassing their networks. It shall also be noted that the entry fee paid by the NLDs and ILDs were much higher than those paid by the ISPs, which led to the above lobbying against Internet telephony. Restrictions on Internet telephony were supposed to provide a level playing field between the various operators who had differing licence conditions.

However, newer protocols such as the Internet protocol version 6 (IPv6), Session Initiation Protocol (SIP) and Multi-protocol Label Switching (MPLS) have improved the quality of service of Internet telephony calls. Details of these technologies are presented in Box 6.4.

Realizing the above, the TRAI in its 2007 recommendations revised the Internet telephony device definitions covering H.323/SIP terminal to include any device/adapter conforming to standards of international agencies like ITU, IETF, and so on (TRAI, 2007).

Box 6.5 gives details of the regulatory directives on Internet telephony (Sridhar, 4 March 2009).

Grey Market in Internet Telephony

The restricted form of Internet telephony encouraged a grey market for related services.

As discussed in previous chapters, IUC regulations imposed ADC on all incoming ILD calls terminating on landline/mobile phone in India. This ADC was as high as Rs 4.25 in 2003 (TRAI, 2003). It was found that some of the ILDOs carried their calls through the Internet till the destination and terminated them as local calls (after masking the originating caller line identification), thus reducing their ADC to be paid to the terminating carrier (refer to Sridhar, 14 October 2004 for details).

Since it is difficult to monitor the deployment of media gateways, it was found that Internet telephony calls from abroad were also being illegally terminated through an IP private branch exchange (PBX) to the PSTN/PLMN (TRAI, 10 May 2007). This again violated the regulations on Internet telephony and also provided a mechanism to evade ADC.

Box 6.4   Advances in Internet Telephony

Session Initiation Protocol

Session Initiation Protocol (SIP), is an open-signalling application layer protocol for establishing any kind of real-time communication session. The communication session can involve voice, video, or instant messaging (IM), *and* can take place on one of many devices that people use for communicating: laptop computer, personal digital assistant (PDA), cell phone, IM client, IP phone, and so on. The SIP has been developed in the *Internet Engineering Task Force (IETF)* by common participation from various vendors, and the specifications are presented in the Request for Comments (RFC) 3261. The SIP introduces a new model for communications through its support of *presence*. Presence enables you to locate a user and determine his willingness and ability to participate in a session, even before you initiate communication. This information, reflected across multiple devices such as IP phones, cell phones, and IM clients, makes communication simple and efficient by helping you to reach the right person at the right time, on the right device. A SIP *address of record (AOR)* provides one unifying identifier (for example, SIP:CoolestProf@mdi.ac.in) that can be mapped across multiple devices and media types. User agents (UAs) are applications in SIP endpoints (such as a SIP phone, cell phone, PDA, or workstation) that interface between the user and the SIP network. A UA can act as either a client or a server. When sending SIP messages, the UA acts as a *user agent client (UAC)*, and when receiving messages, it acts as a *user agent server (UAS)*. The SIP trunks enable enterprises to carry their voice data over a pure IP connection to carrier clouds, rather than through separate circuits as is done traditionally. The SIP addresses can be converted into phone numbers, email addresses or IM addresses or vice versa through databases such as Lightweight Directory Access Protocol (LDAP) or ENUM. ENUM is a protocol defined by the Internet Engineering Task Force's (IETF) Telephone Number Mapping Working Group. ENUM defines a domain name system (DNS) based architecture and protocols for mapping a telephone number to a uniform resource identifier (URI) which can be used to contact a resource associated with that number on the Internet. For details on SIP, refer to Gregory (2006).

Internet Protocol Version 6

Each phone in the world, landline or mobile, is given a unique phone number, dialling which we communicate with our friends and relatives. Similarly, each computer you are using, either at home/work or at a cyber café to send email or browse the World Wide Web, is connected to the Internet and is assigned a unique address referred to as the 'Internet protocol (IP) Address'. Internet Protocol, which assigns unique addresses to every computer on the Internet and enables them to communicate, got a new avatar in IP version 6 (IPv6), the next-generation protocol.

The current version of IP (Version 4) has served us well until now with virtually no change, since it was proposed in 1981. However, the address space provided by IPv4 has not been able accommodate the exponential growth of the Internet witnessed in recent times. Though most of the educational institutes, network service providers and medium/large organizations in the US and Europe have enough addresses to distribute to their network devices, it was felt that worldwide there will be shortages in address space. The problem is acute especially in developing countries such as India, which entered the Internet era only recently. For example, most of the educational institutes in developed countries have a class of addresses (referred to as Class B addresses) that can connect up to about 65,000 network devices in their networks. In India even ISPs have Class C addresses that allow them to allocate only a maximum of about 250 unique IP addresses to their residential and corporate customers.

To support hundreds of computers in the campus using such a limited number of addresses, organizations have to resort to novel methods. There are classes of addresses referred to as 'private addresses', reserved for use by organizations in their private network. Private address space allows organizations to use these addresses internally, within their Intranet. These addresses should never be visible to external networks. Any message that goes out of the Intranet to the public Internet has to be masked with one of the few public IP addresses doled out by the ISPs. This translation is typically done by software known as Network Address Translator (NAT). The NAT is normally deployed on edge network devices such as the firewall or router. While we can live with NAT and private addresses, network performance and security takes a beating. However, almost all organizations in India use the above process to overcome the scarcity of the Internet address space.

Recognizing the exhaustion of the address space, the IETF called for a proposal for implementing next generation of Internet addresses in 1992. In 1996, 6Bone, a test bed for the next generation of IP, namely IPv6, was constructed; and in 1998 IETF accepted IPv6 as a draft standard. IPv6 allows more than a hundred million trillion (1 followed by twenty zeros!) addresses to be allocated for each square inch of the earth's surface and should last for quite some time! For the technically inclined the packet format of IPv4 and IPv6 are compared in the following diagram.

However, IPv6 is not being aggressively deployed all over the world, especially in the US. Even though router vendors such as Cisco started rolling out IPv6-compliant products in 2001, the adoption level is still very low. Ironically, the US, the birthplace of the Internet which has snatched up more than 3 billion IP addresses, has been slow to respond in migrating to IPv6. Most of the US organizations that have Class A and B addresses that can provide addresses for 16 million and 65,000 hosts respectively do not feel the need for migrating to IPv6. The main reasons for this reluctance to migrate being the cost and delay associated with the changeover process. The IETF has also been slow in standardizing the tools required for the interoperability of IPv4 and IPv6 networks. Only recently the Department of Defence in the US has made IPv6 as one of the network infrastructure procurement requirement. The momentum to migrate to IPv6 has been better in the Asia Pacific region, thanks to the drought in IPv4 addresses. However, Japan (63 per cent) and South Korea (23 per cent) lead the race in IPv6 deployment. India does not figure at all in IPv6 deployment.

IPv4 Packet				IPv6 Packet			
Bits				Bits			
0	7	1 5	2 3	3 1	0	7	1 5	2 3	3 1
V	IHL	TOS	Total Length	V	Class	Flow Label
Identification		F	Fragment Offset	Payload Length	Next Hdr	Hop Limit
TTL	Protocol	Header Checksum	Source Address (16 bytes)
Source Address		
Destination Address			Destination Address (16 bytes)
Option (0–40 bytes)			Payload
Payload			

Migrating to IPv6 has a lot of advantages including improved network security and network performance; promotion of mobile and ubiquitous computing by assigning addresses to laptops, cell phones and PDAs; and improved quality of communication. While developed countries can wait for the migration, India and other developing nations cannot afford to do so (refer to Sridhar, 22 October 2003 for more details).

Realizing this, the TRAI released a consultation paper on IPv6 in August 2005 (TRAI, 2005). Some of the Universities and Institutes in India along with Department of Information Technology (DIT) have been doing research on IPv6 deployment. For example, the Engineering and Research Network (ERNET) of DIT in association with the Indian Institute of Technology (IIT), Kanpur has taken up a project of setting up IPv6 test bed at a few locations in the country. However, the ISPs in India have not been aggressively deploying IPv6. Deploying IPv6 will remove the need for NAT; provide address space for mobile devices, corporate network elements and for retail customers; improve routing efficiency; improve quality of service for different kinds of services including Internet telephony; and increase security of networks.

Box 6.5   The Regulatory Dilemma on Internet Telephony

By sending the recommendations on unrestricted Internet telephony back to the TRAI, the Telecom Commission has again vacillated in adopting a superior technology due to possible lobbying by the incumbent telecom companies.

Internet telephony (also referred to as VoIP) is a technology that allows the user to make telephone calls using a broadband Internet connection instead of on a regular landline or mobile. The call can be computer to computer, phone to phone and computer to phone. While the first two methods have not been worrisome for the regulators, the third form has always been controversial all over the world. India opened up a restricted form of Internet telephony, allowing the first two modes and a restricted version of the third (not to terminate VoIP calls on landline or mobile in India) way back on 1 April 2002. Though unrestricted VoIP was allowed for UASPs and CMSPs in March 2006, the telecom companies never promoted this service for the fear of losing their lucrative voice service revenue. With all good intentions, the TRAI recommended in August 2008 that even ISPs could provide unrestricted VoIP, as it would bring down the price of ILD and NLD calls.

In VoIP, real-time voice conversations are converted into data packets and delivered over the unmanaged Internet, which uses packet-switching technologies compared to resource-hungry circuit-switching deployed in the traditional PSTN. In a PC-phone VoIP call, the ISP carries the call from the PC all the way through the Internet cloud, to the Internet telephony gateway near the destination. After performing the required address/content translation at the gateway, hands over the call to the terminating mobile/landline network. This mode of transmission bypasses the traditional NLD/ILD carrier network and hence the resistance by the telecom companies.

There are arguments for and against regulating Internet telephony. Google and Skype alike strongly oppose any regulation on the VoIP calls, and claim that VoIP is just one of the many services (such as chatting, browsing, email) that a VoIP provider/ISP offers to the customers. Hence it shall be considered as an 'information service' and not as traditional POTS provided by the telecom companies. Moreover, when the user makes a VoIP call, it is difficult to determine the exact location of the origination/ destination of the call (for example, location of the PC, location of the gateway) and hence it is not possible to bring it under interconnection regulation for levying termination access charges as is currently practiced in a PSTN call.

However, the proponents argue that technology advancements have made quality of the VoIP calls a close substitute for PSTN calls and hence should be treated the same way as PSTN calls. Moreover, they argue that with the low-cost VoIP adapters that can be plugged on to connect the inexpensive analogue phone to the broadband line, VoIP is becoming widely accessible much like the PSTN and hence should be regulated the same way, including the mandatory emergency numbering, and contribution to the USO fund.

However, in India, broadband connectivity is poor. The operators have just started their 3G mobile broadband service. The broadband wireless access service is yet to be rolled out by the licensees. The VoIP with its low call costs is likely to give a boost to broadband services. Removing restrictions on VoIP will also curb any illegal setting up of the gateways, especially for routing the incoming ILD calls.

While the telecom companies were complaining that VoIP would give undue advantage to the ISPs who paid very low licence fees compared to them, creation of a level playing field is in process. The DoT implemented in 2006, a licence fee of 6 per cent on revenue (including contribution to the USO fund) for the Internet telephony service offered by the ISPs. This is in line with the licence fee for NLD and ILD services. Hence at this stage there is no strong reason for not allowing unrestricted VoIP. In fact by allowing VoIP, competition in NLD and ILD services is likely to increase, thus bringing in much lower prices for consumers. The wire-line access providers tend to gain as VoIP generates revenue for broadband usage.

However, regulators all over the world are keeping a close watch on the evolution of VoIP. Indications are that as VoIP matures to become a close substitute for PSTN calls, regulation of the same may be in the offing as is currently being implemented in the US.

Before it becomes too late, let us benefit from this technology innovation, which has really made 'death of time, death of distance' possible in telecommunications.

Realizing this the TRAI recommended, and later the DoT implemented in March 2006, a licence fee of 6 per cent of the AGR for the Internet telephony service. This is in line with the licence fee for NLD and ILD services. The reason behind this is that the quality of Internet telephony calls have improved considerably thanks to technology advances such as SIP, IPv6 and other mechanisms and hence is comparable to that provided by ILDOs and NLDOs over circuit-switched networks. Further, a service tax was also levied on Internet telephony call charges much like that on calls over the PSTN/PLMN. However, these are applicable only for Indian ITSPs. They cannot be forced on foreign ITSPs such as Skype who have their media gateways outside India. The Indian ITSPs alleged that since foreign ITSPs do not pay (or do not have to pay) licence fee and service tax, it was a revenue loss to the government as well as discrimination against them (Sridhar, 19 December 2006).

In a note, the government, in its nationalistic fervour to protect the security of the country, was also planning to monitor Internet telephony services used by ITES and BPO companies, especially to clamp down on the services of foreign providers such as Net2Phone, Diaplad, or Skype. Cheered by the Indian ITSPs, the government thought that this move would minimize security threats and contain the evasion of service tax and revenue share by the foreign operators. Even if the government is successful in monitoring the usage of foreign providers in business organizations, what prevents a user from downloading Gizmo Project (gizmoproject.com), a strong competitor to Skype, at his/her home PC to make Internet telephony calls?

Internet telephony is just another application on the Internet. There are companies such as Bittorrent.com and Narrowstep.com that provide television through the Internet. Radio stations such as apnaradio.com broadcast their programmes through the Internet. Will these be regulated under the archaic cable TV regulation Act? Internet Messenger (IM) services offered by Yahoo!, MSN, Google and India Times are equally pervasive to provide collaborative communication platforms supporting file transfers, whiteboard, audio and video conferencing features and even mobility. Efforts are underway in the IETF (www.ietf.org) on other applications such as video multicasting, global scheduling and calendaring, and distributed authoring over the Internet.

Does the government add licenses or block each and every Internet application being so discovered? This is against the egalitarian philosophy of the Internet. Any policy the government and the regulator develop should be technology-neutral so that it stands the test of time.

Unrestricted Internet Telephony

Realizing the potential of Internet telephony services, the government allowed access providers such as BTOs, UASPs, and CMSPs to provide unrestricted Internet telephony in March 2006. However, until September 2007, none of the access providers including basic telecom service providers provided Internet telephony service. Voice is still the main revenue-earning service for access providers. If the access providers provide Internet telephony, the consumers expect much lower prices especially for NLD and ILD calls carried through the Internet. Though costs for providing the Internet telephony service is also comparatively much less for the telecom companies, they are reluctant to provide this service for the fear of declining ARPU. It was only in September 2007, that the government operator MTNL announced the provisioning of unrestricted Internet telephony for its Mumbai and Delhi subscribers. This was due to the declining ARPU and subscriber base for its core fixed-line telecom service.

Realizing the advantages of Internet telephony, the TRAI made its recommendations to allow unrestricted Internet telephony to be offered by ISPs as well (TRAI, August 2008). It also argued for better interconnection between the ITSP and NLD operators to enable connectivity between the Internet and the PSTN/PLMN. Prior to these recommendations, the TRAI also levied 6 per cent of the revenue from Internet telephony services as licence fees which is similar to the annual licence fee paid by the NLDOs and ILDOs. However, in March 2008, the Telecom Commission of the DoT buckled under the lobbying pressure of the incumbent telecom companies and sent back the recommendations to the TRAI for further review (Sridhar, 4 March 2009).

Virtual Private Networks

A Virtual Private Network (VPN) is a private network that makes use of the public telecom infrastructure, thus maintaining the required security and quality of service. The VPNs can be contrasted with owned or leased private lines which are end-to-end and are dedicated to one customer. The objective of VPN is to provide secure guaranteed quality of service over the public network at a much less cost compared to leased/owned private network. Thus the VPN provides a cost-effective alternative to the private leased-line networks offered by NLDOs, basic service operators and infrastructure providers. Though various protocols in the computer communication hierarchy can be used to provide VPN services, the Internet protocol (IP) is the most widely used. The open platform of the Internet allows the setting up of secure virtual end-to-end links of required capacity, thus providing VPN services.

The ITU has defined the various types of VPN services, details of which are given in Box 6.6.

The VPN services offered by the ISPs have been one of the contentious issues. The DoT issued guidelines on 16 December 2004 that allowed provision of VPN services by ISPs with an annual licence fee of 8 per cent of the AGR generated under the licence and a one-time entry fee of Rs 10 crores, Rs 2 crores and Rs 1 crore for Category A, B, and C respectively. However, the Internet Service Providers Association of India (ISPAI) filed a petition against the imposed fee structure before the TDSAT.

Subsequently, the TRAI released a consultation paper on the VPN service provided by the ISPs in 2005. After extensive discussions with the stakeholders, the TRAI concluded that the Layer 2 and Layer 3 VPN services are different from Layer 1 VPN services which are akin to the

Box 6.6   Different Types of Virtual Private Network Services

The figure below illustrates different types of VPN services.

Layer 1 Virtual Private Network Service

In a Layer 1 VPN service the CPE is connected to the network provider via one or more links, where each link may consist of one or more channels or sub-channels (time division or frequency division multiplexed). The CPE is connected to the service provider's node through a port. End-to-end connectivity is maintained through an interconnected set of such links. This is equivalent to private leased line circuits connecting the CPEs at the origin and destination, providing complete privacy and security.

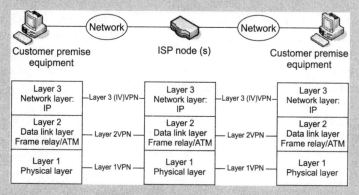

Layer 2 Virtual Private Network Service

In a Layer 2 VPN, the circuits are connected to each other at the data link layer level. The data link layer could be frame relay service (FRS), asynchronous transfer mode (ATM) or point to point (PPP).

Layer 3 Virtual Private Network Service

In this type of service, the CPE is connected at the network layer to the service provider's node. Peer to peer connection between the originating and terminating CPEs is established at the network layer level. This necessitates that the CPEs are Layer 3 devices such as a router. The most widely used Layer 3 protocol is IP. Hence, by default, most of the Layer 3 VPN services are IP VPN services.

leased line service provided by BTOs and NLDOs. Also, it was noted that Layer 3 VPN services are an integral part of the Internet service and hence needs to be regulated lightly with the benefits accruing to subscribers. The VPN can be thought of as an application (much similar to Internet telephony or Voice over IP) that provides secure and quality transmission of information over the public Internet. Security and quality of service are normally deployed at the CPE. The Internet acts as a 'stupid network' just passing information as bits without having to worry about whether it is voice, data or video, secure or not. This property of the Internet provides scalability as new applications evolve.

The cost advantage of VPN over the private leased line network comes primarily because it offers services over a public network. A new licence that required upfront entry fee for provisioning of VPN services would erode the cost advantage being offered by VPN and kill the deployment of this superior technology. However, there was a pressure on the government for considering licensing of VPN services due to the fear that it took away business from other leased line operators such as NLDOs (predominantly government-owned BSNL). However, it must be noted that most of ISPs still use the existing leased line infrastructure provided by NLDOs or infrastructure providers. Hence part of the revenue earned by ISPs goes to these facility-based operators, which in turn enhances their revenue.

In line with the above arguments, the TRAI in its recommendations on 16 August 2005, proposed nil annual licence fee for both Layer 2 and Layer 3 VPN services. An entry fee of Rs 30 lakhs was proposed for Layer 2 VPN service (TRAI, August 2005).

However, due to the successful lobbying by the NLDOs, the government in December 2005, refuted TRAI's recommendations and disallowed ISPs to provide VPN services. If ISPs were to provide VPN services, they were required to migrate to NLD/ILD licenses. It must be noted that by this time, the NLD/ILD entry fees were brought down to Rs 2.5 crores. Most of the ISPs including pure-play ISPs such as Net4India and Sify migrated to the NLD licence so that they could legally provide VPN services. Apart from the entry fees, the ISPs providing VPN services also have to pay 6 per cent of AGR as annual licence fee.

In the above case, there are certain there are certain lessons to be learnt by ISPs as well. ISPs have learnt over a period of time that it is not enough to get just the licence. Even though some 437 ISP licenses were issued by December 2000, only a handful of ISPs have significant business and presence in the market. In fact the low licence fee has also been a curse for ISPs. Most of the basic, unified access, and Multi Service Operators obtained ISP licence by paying the token fee and are able to bundle Internet service with their other services. Hence stand-alone ISPs face intense competition.

Apart from price, customers expect good quality of service. If ISPs have to build a robust business model around VPNs, they better offer services adequate to meet the requirements of users, especially in terms of response time, and security. ISPs often complain that 60–80 per cent of their VPN/ Internet service revenue go towards the cost of leased lines taken from facility based operators and hence cut in to their bottom-line. With competition in the domestic bandwidth from IPs, NLDOs, Basic and unified access operators, the leased line prices are expected to drop further. The existing licence allows ISPs to establish their own transmission links within their service area if such capacity is not available from authorized agency. In fact most of the ISPs do provide last mile linkage to their network Points of Presence with their own transmission facilities. The regulator and the government shall modify this interconnect agreement suitably so that ISPs will also be able to set up their own networks in case they are bullied by the incumbent leased line providers.

Internet Exchange

The Internet is a network of networks as illustrated in Figure 6.1. The response time for access to an Internet resource (for example, a web page) depends on how well the different networks are interconnected to each other. There are two ways by which an ISP's network is interconnected to other networks, (i) via a common Internet exchange (ii) by a peering connection with other ISP networks or with the national Internet backbone. Option (ii) is possibly infeasible with so many ISPs in the market. However, most of Category B and Category C ISPs are small and hence interconnect with upstream Category A ISPs to take their traffic.

The TRAI had set up a task force in 2002 to examine the slow growth of the Internet in the country. Based on the recommendations of the task force, the government set up the National Internet Exchange of India (NIXI) as a non-profit company with initial grant from the Department of Information Technology (TRAI, April 2007). The objective was to have a common point of interconnect between various ISPs in the country so that (i) the domestic traffic is routed within the internal networks of the country, thus improving response time and (ii) the local ISPs do not have to pay international traffic charges for routing traffic within the country which otherwise could have been routed through the international links. The NIXI nodes were set up in the four metros of Delhi (Noida), Chennai, Kolkata, and Mumbai.

However, until 2007, the use of the NIXI by the ISPs was poor due to the following reasons:

1. The ISPs (especially Category B and C ISPs) were not able to afford the high cost of leasing lines from their network PoP to the NIXI nodes located at the four metros.
2. Most of the ISPs did not have the Autonomous Systems (AS) number which was mandatory for connectivity to the NIXI.
   a. Getting AS number from APNIC was relatively expensive (for non-members: US$ 100 one-time fee and US$ 50 per year for AS number; for members: member fee charges of US$ 1,250).
3. Due to non-advertisement of routes by the ISPs who are connected to the NIXI nodes, there were poor interconnections. New ISPs could not see any benefit by connecting to NIXI nodes.

Realizing the poor utilization of the NIXI nodes, the TRAI recommended the following in April 2007 (TRAI, April 2007):
1. All the ISPs or their upstream providers should either be connected at all NIXI nodes or to international Internet bandwidth providers through domestic peering link. All the ISPs providing international Internet bandwidth should be connected at all the four nodes of NIXI. This ensured that domestic traffic is treated differently by the international Internet bandwidth providers, at the same time allowing some flexibility to the ISPs to connect. This also eliminated the need for all the ISPs to get an AS number to be (indirectly) connected to the NIXI nodes.
2. All the ISPs should announce and accept all their routes at NIXI nodes.

However, it must be pointed out that the TRAI did not find it feasible either to interconnect the four NIXI nodes or to set up additional nodes, especially in the state capitals.

Without augmenting the NIXI infrastructure, the efficient utilization of the NIXI nodes cannot be guaranteed. The national Internet backbone which is being set up by the Department of Information Technology can be used as the backbone for connecting the NIXI nodes and for setting up additional NIXI nodes. The NIXI can generate revenue and profit as well, from the ISPs for transferring the traffic. Continuing with inadequate infrastructure is not likely to solve the problem of ineffective utilization (for details on TRAI's recommendations, refer to TRAI, April 2007).

Domain Names

As discussed, every device connected to the Internet has to be provided with a unique IP address. The IP address is difficult to remember for an Internet user. Hence the *domain* concept is used to map the IP address of the device into a mnemonic format that can be easily remembered. Domain refers to a group of networks that are under the administrative control of a single entity, such as a company or a government agency. Domains are organized hierarchically, the details of which are provided in Box 6.7. The NIXI (http://www.nixi.in) is the official .in domain name registry of India.

The structure of the .in domain is given in Figure 6.5 (NIXI, 2007).

Most of the registrants in India preferred to use their domains registered under generic TLDs such as .com or .org until recently. With the NIXI lowering the prices of domain name registration under the .in domain, the registration under .in rose from about 300 per day to 3,000 per day in 2007, thus totalling about 400,000. Efforts are underway at NIXI to develop India-specific IDNs in five to six Indian languages (namely Hindi, Tamil, Sanskrit, Bengali, Punjabi, and some others).

Cybersquatting and Domain Name Resolution

Cybersquatting is the practice by which a person or legal entity books up the trademark, business name or service mark of another as his own domain name for the purpose of holding on to it and thereafter selling the same domain name to the other person for valuable premium and consideration (Cyber Law India, 2008). The owner of the original trademark/business

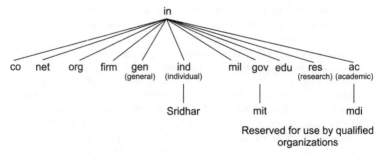

Figure 6.5   The .in Domain Name Tree

Box 6.7   Domain Name System

Domain refers to a group of networks that are under the administrative control of a single entity, such as a company or a government agency. Domains are organized hierarchically, so that a given domain consists of different subordinate domains. The following figure shows an example domain in the domain hierarchy.

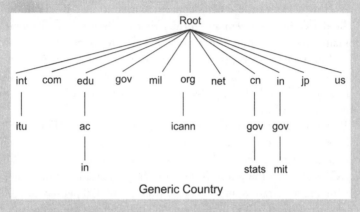

Domain Name Tree

The Internet Corporation for Assigned Names and Numbers (ICANN) is responsible for the global coordination of domain names and for allocating the Top-level Domains (TLDs). The TLDs are of two types:

(i) Generic domains (gTLDs) refer to domains associated with international organizations (.int), educational institutions (.edu), commercial organizations (.com), government agencies (.gov), network service providers (.net), military agencies (.mil), and professional organizations (.org). A new set of domain names associated with businesses (.biz), museums (.museum), air transport industry (.aero), cooperative associations (.coop), information (.info), mobile products and services (.mobi) have been created by the ICANN for catering to the associated beneficiaries (for a full list of all TLDs, refer to http://www.icann.org).

(ii) Country domains (ccTLDs) which are specified for each country such as the .us for the United States, .in for India and .cn for China.

Efforts are underway at the ICANN to internationalize domain names, representing the domain name entirely in scripts other than the familiar Latin characters that appear in current top-level labels. The internationalized domain names (IDNs) made the domain name label as it is displayed and viewed by the end-user different from that transmitted in the Domain Name System

(DNS). The IDN uses the following labels and Unicode representation of the IDN (ICANN, 2008):

- The A-label is what is transmitted in the DNS protocol. This is the American Standard Code for Information Interchange (ASCII) compatible form of an Internationalizing Domain Names in Applications (IDNA) string; for example 'xn--11b5bs1di'.
- The U-label is what should be displayed to the user and is the representation of the IDN in Unicode; for example '*pariksha*' ('test' version in Hindi, Devanagari script).
- Lastly, the Letters, Digits, Hyphen (LDH) label strictly refers to an all-ASCII label that obeys the 'hostname' conventions and that is not an IDN; for example 'icann' in the domain name 'icann.org'.

The domain names are converted into the corresponding IP address of the associated host using domain name servers through recursive queries (Tanenbaum, 1996).

name or service mark can dispute the wrong holding of the domain name. The World Intellectual Property Organization (WIPO) under the auspices of the United Nations formulated the Uniform Domain Name Dispute Resolution Policy (UDRP) in December 1999. By 2006, 10,177 UDRP or UDRP-based cases (gTLD and ccTLD) have been filed with the WIPO dispute resolution centre, covering 18,760 separate domain names (WIPO, 2008). The WIPO dispute resolution procedure served a wide range of users, ranging from well-known brands, to smaller enterprises and organizations, as well as individuals. They covered categories including luxury items, famous persons, entertainment, hospitality, sports, gambling, and pharmaceuticals. In addition, charitable organizations and educational institutions were involved.

Under the UDRP of WIPO, the first Indian organization to win a legal battle and get back the cybersquatted domain name of *theeconomictimes. com* and *thetimesofIndia.com* was Bennett Coleman & Company Limited in March 2000 (for details of the case, refer to Box 6.8).

In India, the NIXI resolves domain name disputes involving the .in domain. The .in Domain Name Dispute Resolution Policy was formulated in 2005 in line with internationally accepted guidelines, and with the relevant provisions of the Indian IT Act 2000. Any person who considers that a registered domain name conflicts with his legitimate rights or interests may file a complaint to the .IN Registry on the following premises:

Box 6.8   Cybersquatting Case of Bennett Coleman & Company Limited

The Economic Times is the nation's most prestigious and widely read financial newspaper which has set benchmarks for the entire financial newspaper industry. On 11 March 2000 panellist W.R. Cornish of the WIPO Arbitration and Mediation Centre delivered the administrative panel decision in the cases entitled 'Bennett Coleman & Co. Ltd. vs. Steven S. Lalwani' and 'Bennett Coleman & Co. Ltd. vs. Long Distance Telephone Company'. In both the cases, Bennett Coleman & Co. Ltd. was the complainant before the WIPO Arbitration and Mediation Centre. The facts arising out of the said issue were that Bennett Coleman and Co. Ltd., being the complainant, publishes *The Economic Times*, which had an average daily circulation of 350,100, and *The Times of India*, which had a daily circulation of 2,522,488. Further, Bennett Coleman & Co. is engaged in certain supplemental activities in the entertainment industry, using the brand 'Times'. Since 1996, the complainant has held the domain names, www.economictimes.com and www.timesofindia. com, using them for the electronic publication of their respective newspapers.

The complainant had registered in India the mark, 'The Economic Times' for newspapers, journals, magazines, books and other literary works on 28 March 1973, and the mark, 'The Times of India' for the same products on 30 July 1943. However, in 1998, Steven S. Lalwani, 16 Victoria Terrace, Upper Montclair, NJ 07043, USA registered the domain name www.theeconomictimes.com with Network Solutions. It may be pertinent to note that the domain name www. thetimesofindia.com was also registered with Network Solutions by the Long Distance Telephone Company, having the same address of Steven S. Lalwani, being 16 Victoria Terrace, Upper Montclair, NJ 07043, USA. The said domain names were duly registered by Network Solutions, the then sole registrar for TLDs as the same were available to be registered on a first-come-first-served basis. Needless to mention, Network Solutions does not prescribe any restrictions on the registration of the domain names.

Thereafter, Steven S. Lalwani and Long Distance Telephone Co., being a front of Lalwani, built up their websites on the two domain names which redirected Internet users and netizens to www.indiaheadlines.com which provides India-related news and articles. The effect of this was that any netizen legitimately wanting to go the site of *The Economic Times* and *The Times of India* would type the said names in his browser and he would be redirected to the site of www.indiaheadlines.com. This redirectional exercise adopted by Lalwani caused tremendous harm and damage to *The Economic Times* and *The Times of India* apart from misleading bonafide and genuine netizens.

All this continued in 1998 and 1999 and despite numerous efforts, no wise counsel prevailed upon Lalwani to desist from doing his redirectional work. This was also the time when the practice of cybersquatting was most prevalent.

By means of a detailed and well-reasoned judgement, the WIPO Arbitration and Mediation Centre delivered its ruling that the respondent's domain name www.theeconomictimes.com should be transferred to the complainant Bennett Coleman & Co. Ltd. Thus, *The Economic Times* created history in the annals of cyber law in India by being the first Indian domain name which was directed to be transferred in the name of complainant under the new ICANN domain name dispute policy. In the connected case, WIPO further directed that the domain name www.thetimesofindia.com be also transferred to Bennett Coleman & Co Ltd.

The importance of the domain name victory of *The Economic Times* in Indian cyber law is because the said judgement laid down some important principles of cyber law relating to domain names. These concrete principles will further guide the development of the law against cybersquatting.

(1) The judgement creates a precedent by establishing the principles that daily usage of newspaper titles/marks in hardcopy and electronic publication leads to a substantial reputation which cannot be allowed to be hijacked in cyberspace by a squatter.

(2) The decision grants judicial recognition to the principle of presumption at the time of the registration of the domain name. The moment anyone registers a domain name which has the trademark or other mark of any other legal entity, it shall be presumed that the said person must have been fully aware of the said mark at the time when he applied for the registration of the domain name. Surely, such a principle does have some exceptions but, broadly speaking, the presumption test applied by the WIPO Centre will serve the purpose.

(3) Another test given by the WIPO Centre in holding the registration and use of domain names as being in bad faith is the test of the site being a 'postal address' to other sites. The normal corollary of the same is that the website being a 'postal address' to other sites would be presumed to be doing so in order to take advantage of the very considerable reputation of the domain names in dispute by misleading Internet users into believing that the site in question is associated with the site of the famous mark/trade name of another.

At this juncture it shall be pertinent to examine the reasoning given in the historic judgement. The first issue raised by the respondents was that the WIPO Arbitration and Mediation Centre has no jurisdiction under the ICANN Policy and Rules to consider the said cases because the domain names in contention were allotted before the dispute settlement procedure was brought into effect. The WIPO Centre refuted this argument of the respondents by holding that the registration agreement of a domain name holder applies not only to initial registration but also to the maintenance and renewal of domain names which, in turn, provides for the incorporation of terms and conditions from time to time and as such the same admits the introduction of the UDRP duly approved by ICANN and associated rules and supplementary rules into existing agreements. As the WIPO Centre's panel was duly constituted in accordance with the said policy, the objection relating to the lack of jurisdiction was rightly overruled.

On merits, the WIPO Centre's panel found that the complainant's essential complaint was based upon the large circulations of its financial newspaper *The Economic Times* and its general newspaper *The Times of India*. It was also contended by the complainant that the respondent had no right or legitimate interest in respect of the domain names which were registered in bad faith in order to attract, for commercial gain, Internet users by creating confusion in their minds. The respondent mainly contended that the complainant had no relevant trademarks registered in the US and that there is no likelihood of confusion between the parties' respective sites and the respondents have built up their brand name recognition, awareness and goodwill under the said domain names.

Source: Adapted from Cyber Law India, 2008

1. the Registrant's domain name is identical or confusingly similar to a name, trademark or service mark in which the Complainant has rights;
2. the Registrant has no rights or legitimate interests in respect of the domain name; and
3. the Registrant's domain name has been registered or is being used in bad faith.

The registrant is required to submit to a mandatory arbitration proceeding in the event that a complainant files a complaint to the .IN Registry, in compliance with this policy and rules thereunder. The .IN Registry shall appoint an arbitrator out of the list of arbitrators maintained by the Registry. The arbitrator shall conduct the arbitration proceedings in accordance with the Arbitration and Conciliation Act 1996 as amended from time to time and also in accordance with this policy and rules provided thereunder (NIXI, 2007).

Future of Internet Services Regulation

The Internet services in India started with light-touch regulation. The licence fee was minimal and rollout obligations were not stringent. This led to a large number of companies getting the ISP licence. However, there were restrictions on (i) last-mile access provisioning (until 2004)

(ii) Internet telephony and (iii) VPNs. The ISPs also did not take a proactive role in

1. developing value-added services and providing quality of service to their customers
2. implementing IPv6 and other emerging technologies towards achieving (1).

Hence by May 2007, out of the 389 ISP licensees, only 135 were functionally active. The decline in the number of operational ISPs is shown in Figure 6.6. The competition index as measured by HHI continuously increased from 0.16 in June 2003 to more than 0.25 in December 2006 (TRAI, May 2007). Out of the top twenty ISPs in the country, there were seven integrated service providers (namely the service providers who also provided basic services, cellular mobile services or unified access services in addition to Internet services), twelve Category A ISPs and one Category B ISP. The top seven ISPs had more than 90 per cent of the market share.

One of the main reasons for the large number of non-functional ISPs was that it took a long time for the ISPs to realize that it was not enough to just get a licence, they also had to provide good service. Most of the pure-play ISPs barring a couple of them did not have their own infrastructure. They had to lease bandwidth from other service providers

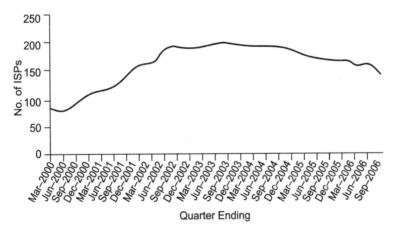

Figure 6.6   Number of Operational Internet Service Providers in India
Source: TRAI, May 2007.

such as BTOs, NLDOs or infrastructure providers. Hence it was not possible for the ISPs to provide guaranteed quality of service on an end-to-end basis (Sridhar, 11 June 2004). On the other hand, the integrated service providers were able to leverage their existing infrastructure to their advantage in providing Internet services bundled with their other services such as voice (refer to Sridhar and Malik, 2007 for details on how a BTO bundled Internet services to garner market share).

The TRAI released its recommendations in May 2007 that seriously looked at the performance of the ISPs in the country, which were subsequently released as guidelines by the DoT (DoT, 2007). Though the subscriber base doubled from 4 million in March 2004 to 8.55 million in December 2006, the revenue did not increase proportionately. The ARPU for a dial-up subscriber decreased from Rs 400 in March 2004 to Rs 200 in March 2005. To tighten the performance level of the ISPs, the TRAI recommended stringent entry and exit conditions.

The new ISP licensing guidelines warrant an entry fee of Rs 20 lakhs and Rs 10 lakhs respectively on new Category A and B ISPs to prevent non-serious players from entering the market. Category C ISP licence was abolished and the existing Category C ISPs were asked to migrate to Category A and B, after fulfilling the financial and rollout conditions appropriately within a certain time period. It was also recommended that all ISPs pay 6 per cent of AGR subject to a minimum of Rs 50,000 and Rs 10,000 for categories A and B respectively. The AGR is calculated on all services including revenue earned from provision of Internet access, value-added services and broadband services. This was done to ensure a level playing field with other telecom operators. The Internet penetration is still very poor in smaller towns and rural areas of the country. These areas require local web content. Smaller niche ISPs, such as Category C ISPs were ideally positioned to provide this service. Though out of 205 Category C licensees only 41 were operational as of December 2006, it is unclear whether the above modification of licensing for Category C ISPs will have enough effect on the provisioning of Internet services in smaller towns and rural areas of the country.

The 2007 ISP guidelines included the following:

1. The ITSP licence was discontinued, however, ISPs could only provide restricted form of Internet telephony service. Internet telephony was still treated as a different service in its scope, nature and kind from real-time voice as offered by other licensed basic, cellular mobile and unified access service licence holders.

2. Direct interconnectivity between two separately licensed ISPs was permitted.
3. Emerging service such as IPTV was permitted for companies having ISP licence and a net worth of Rs 100 crore or more.
4. The ISP licensee was not permitted to have PSTN/PLMN connectivity.

In most of the countries the Internet services are not strictly regulated. The reason being that the services in the Internet keep on evolving and it is difficult for policies and regulation to keep pace. However, in India, the regulation is moving from light-touch to micro-regulation for Internet services. The regulator treats Internet telephony as a different service in its scope from real-time voice services as offered by BSOs, CMSPs, UASLs, NLDOs and ILDOs. Hence it does not allow unrestricted PSTN/ PLMN interconnectivity. However, the ISPs were asked to pay the annual licensing fee (as a percentage of AGR) and service tax as the other service providers. This spells doom for pure-play ISPs. The integrated service providers are in a better position to use the above changes in regulation to become stronger in the Internet service space.

The Net Neutrality Debate

With the development of mobile broadband, the 'Network Neutrality' principle is coming into the limelight and is being argued. Network neutrality is a principle that says those who operate networks which provide an overall benefit to the public good and rely on public property should not use their ownership to confer discriminatory treatment among their customers (Weiss, 2006). Suppose you built a private toll highway and it is the only route from Point A to Point B available for most drivers without off-road vehicles. It is your right to set a toll rate for access to the road, which net neutrality advocates define—perhaps vaguely—as 'fair and reasonable'. It is even your right to charge different rates for different lanes. For example, you could charge a premium to drive in the high-speed lane, which is wider, better maintained, and posts higher speed limits. This is how the broadband network and charges are currently structured. Now suppose Honda signs a deal with you, the highway owner. In exchange for a fee from Honda, you allow only Honda cars in the fastest lane. Neither Hyundais nor Marutis can access the fastest lane at any price. You are now discriminating among your customers, because they do not all have equal access to the service at an open rate. The net neutrality principle is violated. This, net neutrality advocates argue, is what the US telecom companies and cable operators want—and what Internet companies such as Google do not want.

According to network neutrality proponents, the above debates threaten innovation in Internet content and applications in two ways. First, content and application providers' success depends in no small part on the number of customers they can reach. Allowing network providers to exercise greater control over the traffic flowing through their networks would enable those network providers to restrict the number of end-users that any particular content or application provider could reach either by blocking them entirely or by charging content and application providers more in order to reach them. Second, TCP/IP fosters a particular type of innovation. Network neutrality proponents argue that one key to the Internet revolution is the commitment to an architecture in which the pipes through which the data flows are as simple and general as possible and in which all of the intelligence is concentrated in the computers operating at the edge of the network. This architecture frees content and application providers from the need to obtain permission from network providers before deploying their innovations. In other words, TCP/IP promotes innovation by decoupling content and application providers from the network through which that content and those applications travel. Any deviation that creates a tighter integration between the network and the content/applications that the network is carrying would chill innovation by raising the danger that part of the value of any innovations might be captured by the network provider.

As video services and the digital distribution of content over the Internet grow, Internet broadband access providers including AT&T, Verizon, and a number of cable TV companies, have recently demanded additional compensation for carrying digital services. Ed Whitacre, the Chief Executive Officer of AT&T, expressed his company's dislike of existing regulatory structures: 'Now what they would like to do is use my pipes free, but I ain't going to let them do that because we have spent this capital and we have to have a return on it'. The claim that consumers, content providers, or applications' providers use the Internet for free is certainly incorrect. Currently, users pay ISPs for access to the Internet. Similarly, ISPs pay fees to Internet backbones for access to the Internet. The ISPs pay per month for a virtual 'pipe' of a certain bandwidth, according to their expected use. When digital content (or information packets of any service) is downloaded by Consumer A from Provider B, both A and B pay. A pays his ISP through his monthly subscription, and B pays similarly. In turn, ISPs pay their respective backbones through their monthly subscriptions. Unlike a traditional telephone call arrangement in which only the calling party pays, Internet backbones collect from both sides of a communication. So, what change would AT&T's CEO like to see in the pricing and industry structure? He desires the abolition of 'net neutrality', the regime that does not distinguish in terms of price between bits or information packets according to the services that they provide, and additionally fails to distinguish price based on the identities of the uploader and downloader. This pricing regime has prevailed since the inception of the commercial Internet. Presently, an information packet used for VOIPs, email, images, or video is priced equally as a part of the large number of packets that correspond to the subscription services of the originating and terminating ISPs. In addition to content neutrality, there is no distinction made according to the identities of the uploader and downloader. AT&T, Verizon, and cable Internet access providers would like to abolish the regime of 'net neutrality' and in its place substitute a pricing schedule that charges both the final customer for his or her basic transmission service and the transmission's originating party (such as Google, for example) for the provision of content. An access network, for example AT&T, wants to charge fees to an originating party even when the originating party does not connect to the Internet using AT&T and therefore does not have any contractual relationship with AT&T. Access network operators have also reserved the right to charge differently based on the identity of the provider even for the same type of packets; for example, an ISP may charge Google more than Yahoo for the same transmission. The proposed Internet model, without 'net neutrality', would more closely mirror the traditional pre-Internet. The classic case is the development of mobile broadband applications that work on smartphones such as Apple's iPhone and is given in Box 6.9 (Venkatesh and Sridhar, 14 September 2009).

In the FCC rulings in December 2010, net neutrality is addressed from the point of view of two classes of services: (i) Fixed wire-line and (ii) Wireless broadband service. The three important rules are given below:

1. Rule 1: Service providers should be transparent about management of their networks and services
2. Rule 2: Blocking of Internet traffic

Box 6.9   Let the Traffic Flow

On 14 July 2009, Apple announced that the number of downloaded applications from their App Store exceeded 1.5 billion. In just one short year after the launch of the online application store the figures unequivocally suggest that this was one of the most successful projects in the mobile industry history. At times when the mobile phone market was in recessionary slump, the availability of more than 65,000 applications was the single most important factor responsible for the very successful adoption of Apple's iPhone. 'Smartphones' like the iPhone have also led to the increasing adoption of wireless broadband, with the global mobile broadband subscription touching 225 million, and Asia-Pacific contributing to about 90 million.

Given this rapid spread of the iPhone and mobile broadband, it is surprising that AT&T, the US mobile service provider who bundles iPhone 3G handsets along with its access service has decided to put restrictions on the iPhone applications that can run on its 3G network. Citing the reason that the huge downloads are clogging its network, AT&T banned applications such as Sling Player that can be used to redirect TV signals onto iPhone through the 3G network. This is not the first time that mobile service providers have sought to restrict the services that flow through their networks. Skype, another victim, had to restrict its cheap and almost free Internet telephony application designed for iPhone, to work on the public Wi-Fi network but not on AT&T's 3G wireless data connection. Only this restricted application is available through the App Store.

The proponents of 'network neutrality', comprising mainly Internet geeks, want prohibition against such blockages of content and applications, especially those that are bandwidth-intensive. Even the non-geeks, like economists argue that content and applications would generate usage of access networks. If there are no content or high-bandwidth applications, the 3G or mobile broadband service is of little value to subscribers. In a competitive market, it should not be in the interest of the access service provider such as AT&T to constrain a customer's ability to access particular content or application.

Net neutrality advocates argue that regulators should restrain or constrain the ability of network operators to enter into the production of content or application. Through its bundled offering, AT&T controls the way in which applications are made available to subscribers through its mobile network. Bundling safeguards the walled-garden approach practised by the operators. Until recently, most countries in Europe did not allow bundling due to the above reason, so that subscribers could have enough freedom of choice of content or application.

With the long awaited 3G auctions completed, Indian operators can be much better prepared in the launch of their services through lessons learnt from other markets. The 3G network service is of little value if bandwidth-intensive applications are scarce. It is in the interest of the operators to work with the content and application developers to develop content that will drive usage of their 3G services. A good example of this is NTT DoCoMo of Japan, which demonstrated its widely successful 2G-iMode, and 3G-FOMA services that use a consortium of partners for content development. DoCoMo pioneered the practice of sharing a much larger portion of the service revenue with its partners than what was the prevailing practice.

There are theoretical models that explain why operators should not be charging content providers for gaining access to the network and should share more revenue with content providers. These models evaluate the tradeoffs in the two-sided market where on the one side the subscriber pays to the operator for access and content and on the other side the operator pays a percentage of content revenue to the content provider, and charges the content provider for access to the network. There are strong cross-side network effects between content and the network access service in this market due to the complementarities between the two. A smaller revenue share by the operator to the content provider would increase the marginal cost of the content and hence increase the final price the consumers would pay. This provides an explanation for why value-added services are not picking up pace in India—service providers in India tend to pay a very small percentage of the content access fee to the content providers.

With only 2×5 MHz available for 3G services currently, the operators will need to manage the use of the limited capacity created through 3G very carefully. One way is for the operators to adopt 'access tiering' which allows the operator to charge higher prices for high-priority content providing guaranteed quality of service and quicker response time. The Indian mobile industry has always been very innovative in pricing, as is evident today with pay-for-what-you-use, pulse-a-second, and credit-on-incoming plans. Access tiering will open up new opportunities for operators and content providers to distinguish their 3G service offerings. However, access tiering is strongly opposed by net neutrality geeks, since this would mean that large incumbent content players could sell their content with higher priority thus hogging the bandwidth and depriving smaller upcoming start-ups any share of the market.

With limited landline-based broadband access at about 6.5 million, mobile broadband holds much promise in India. Competition in the market, an appropriate ecosystem promoting collaboration between content/application providers and operators, and innovative pricing schemes could pave the way for the success of mobile broadband in India.

For fixed wire-line networks, operators cannot block any lawful content, services, applications, or devices on their network. The rules will allow fixed broadband providers to charge consumers according to usage, a metered pricing practice already used by some wireless carriers.

Wireless providers are also prohibited from blocking websites, but the rule is slightly more lenient when it comes to blocking applications and services. They are not allowed to block VoIP applications such as Skype and may not block websites in their entirety.

Rule 3: Prohibits fixed wire-line broadband providers from unreasonably discriminating against traffic on their network.

The reason given by the Chairman of FCC for being lenient towards mobile services is that mobile broadband services are still in the earlier stages of adoption. Hence restrictions in the form of network neutrality may prevent deployment and further adoption of wireless broadband and mobile Internet. Immediately following the ruling, Verizon, the largest mobile as well as broadband carrier in the US filed a legal challenge to the net neutrality rules passed by the FCC, arguing that the regulators overstepped their authority.

Broadband and the Internet

Internet and broadband go hand-in-hand. As discussed in a previous section, Internet access through the broadband network is still very limited in India. Figure 6.7 illustrates the broadband and non-broadband Internet connections in India.

Though broadband connections have been growing, the rate of growth is inadequate. The broadband growth has not only been slow but also biased in favour of urban areas. More than 60 per cent broadband subscribers are in the top ten metros and Tier-I cities and more than 75 per cent connections are in the top thirty cities. Just 5 per cent of the broadband connections are in rural areas which are meagre compared to

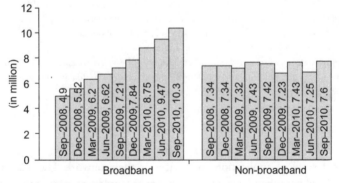

Figure 6.7 Internet Access
Source: TRAI, 2010.

about 31 per cent of the total mobile telephone connections in rural areas. Figure 6.8 gives the broadband penetration for the top thirty-two cities.

Realizing the skewed growth of broadband in the country, the TRAI released a consultation paper on the National Broadband Plan in 2010 and came out with the following salient recommendations (TRAI, 2010):

Seventy-five million broadband connections to be provided by the end of the year 2012 and 160 million broadband connections to be provided by 2014.

1. Towards the above, the national broadband network will be established. This network will be an open-access optical fibre network connecting all habitations with population of 500 and above. This network will be established in two phases. The first phase covering all cities, urban areas and Gram Panchayats will be completed by the year 2012. Phase II will see the extension of the network to all the habitations having a population more than 500, to be completed by the year 2013.

2. This network will be established at a cost of about Rs 60,000 crores. It will be financed by the USO fund and the loan given/guaranteed by the Central Government.

3. Realizing the need for a robust well-connected fibre optic backbone network across the country, it is proposed that a 100 per cent, Central Government-owned, holding company called the National Optical Fibre Agency (NOFA) be formed. The primary objective of such an agency is to plan, install, operate and maintain shared fibre network in

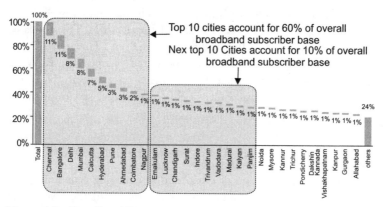

Figure 6.8 Broadband Penetration in Select Cities in India
Source: TRAI, 2010.

the sixty-three Jawaharlal Nehru Urban Renewal Mission (JNURM)-identified cities and provide means to allow any service provider to use the network for giving broadband connections using any technology in the last mile.

4. A State Optical Fibre Agency (SOFA) should be formed in every State with 51 per cent equity held by NOFA and 49 per cent by the respective State Government. The NOFA would be the holding company of all the SOFAs. The main functions of the SOFA shall be:

   a. Plan, install, operate and maintain access aggregation network in the rural areas to connect various access networks deployed by the service providers to the block headquarters.

   b. Plan, install, operate and maintain the backhaul between block headquarters and the district/state headquarters and provide means to connect it to any service provider's backbone network.

   c. Plan, install, operate and maintain shared fibre network in urban areas other than the sixty-three JNURM-identified cities and provide means to allow any service provider to use the network for giving broadband connections using any technology in the last mile.

5. Broadband connection definition as described in the Broadband Policy shall be rewritten as follows:

   'A data connection using any technology that is able to support interactive services including Internet access and support a minimum download speed of 512 Kilobits per second (Kbps).'

   It is to be noted that the upload speed will be at least be half the download speed. This definition of broadband (both wire-line and wireless) given above is effective from 1 January 2011. The stipulated download speed of 2 Mbps will be effective from 1 January 2015.

6. The government may fix and notify the charges for RoW in consultation with the State Governments on a priority basis and ensure time-bound availability of RoW to telecom service providers after due intimation to the agency concerned.

The ecosystem required for the deployment of broadband networks and services is presented in Figure 6.9.

Critical to the adoption of broadband in emerging economies such as India are (i) price of connectivity and services (ii) the diversity of applications that cater to the local needs of the adopters and (iii) the governance mechanisms integrating the interests of various stakeholders as given in Figure 6.9.

For the initial deployment and adoption of the broadband networks and services, it is recommended by both the TRAI (2010) and Jain *et al.* (2010) that the USO fund shall be used for building the network infrastructure. The application development for wireless broadband services is still in its early stages in India. Availability of broadband infrastructure is not sufficient for the adoption of broadband services (see Box 6.10 for a case study of common service centres). Jain *et al.* (2010) also cite local loop unbundling as a possible option to improve broadband infrastructure in the country, details of which have been described in Chapter 2.

The above case clearly points out the need for the ecosystem build-out as presented in Figure 6.9 for large-scale adoption of broadband in the country.

* * *

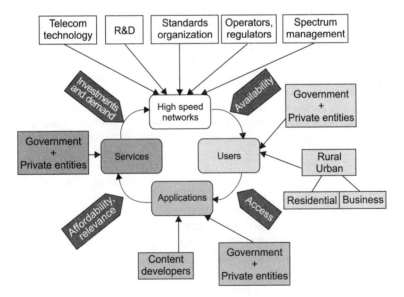

Figure 6.9   The Broadband Ecosystem
Source: Jain *et al.*, 2010.

Box 6.10   Case of Common Service Centres: Infrastructure Alone is not Sufficient

Common service centres (CSCs) were set up to provide high-quality and cost-effective video, voice and data content and services, in various areas of e-governance as well as other private services. The focus was to set up the CSCs in the rural areas in order to make government services accessible to even the most remote village.

The CSC scheme had been approved by the government in September 2006 with an outlay of Rs 5,742 crores over a period of four years. A typical CSC was to be a retail outlet of services that offered solutions based on a combination of ICT infrastructure (PCs, printers, scanners, digital camera, projection systems, tele-medicine equipments, and so on), rural entrepreneurship and market mechanisms. The CSC was established through a bottom-up approach and was customer-centric and a single window for all government to customer (G2C) services and other retail functions.

The CSC scheme had been planned to be implemented in a public–private partnership framework. This model envisaged a three-tier structure consisting of the CSC operator (called village level entrepreneur or VLE), the Service Centre Agency (SCA) that would be responsible for a division of 500–1000 CSCs and a State-designated Agency (SDA) identified by the state government responsible for managing the implementation over the entire state.

Though the CSC project received approval in September 2006 with a two-year plan for rollout, most of the next two years, 2007 and 2008 were consumed in sensitizing the various private players about the project, holding meetings with them and initiating the bidding process. In the year 2008, only ten states had started the rollout of the CSCs, adding up to 14,216 centres across these states.

Though the private sector responded enthusiastically when advertisements were put for SCAs and more than 2,300 organizations from across the country and with diverse backgrounds responded to this advertisement, the implementation till 2010 has not been satisfactory. Out of the hundred thousand centres proposed by the Ministry of Information Technology, about 45,686 centres were pending in June 2010. It is also noteworthy that even though a state had achieved decent rollout, it did not mean that all its centres were fully operational. The initiating idea of the CSCs was to provide a bundle of services, predominantly in the online mode that enabled provision of selected government services to the citizens along with providing other services.

Following are the reasons cited for the poor deployment of CSCs:

(i)   Lack of broadband connectivity and associated IT infrastructure
(ii) Lack of appropriate government to consumer services and
(iii) Lack of support from government officials and availability of skill sets in managing these centres.

Source: Adapted from Jain *et al.*, 2010

This chapter highlights the growth of Internet services and the regulations that prevent the stakeholders from reaping the benefits of the Internet. With the recent regulations, the Internet services in India have moved from light-touch regulation to a more organized regulated industry. With broadband becoming a necessity for economic development, we can only hope that Internet services in the country will pick up in the coming years.

7

Satellite Networks and Services

Solutions for Pan-India Coverage

In the 1970s, VSAT networks emerged as a low-cost satellite-based wide-area network option especially suitable for private data interchange. While VSATs found wide applicability in enterprise networks in the developed countries in the 1980s, the Indian VSAT market started developing in the 1990s when in 1992, the government allowed VSATs to be used for private wide-area data network connectivity.

Very Small Aperture Terminal Technologies

Very small aperture terminals, as the name indicates provide network connectivity, especially across geographically distributed locations, using transponders mounted in the geostationary (GEO) satellites. The GEO satellites hovering at 36,000 km above the earth's surface are relatively stationary with respect to the earth and hence provide an ideal medium to transmit signals across distant locations. Typically a GEO satellite covers about one-third of the earth's surface. The earth stations, also called terminals, are typically located at the customer premise and operate at designated frequencies. The VSAT technologies are detailed in Box 7.1.

Very Small Aperture Terminal Applications

Very Small Aperture Terminal for Voice

The VSAT networks can be used as access networks for providing telecom services especially in remote and rural areas of the country. They

Box 7.1   Technology of Very Small Aperture Terminals

A VSAT system consists of a satellite transponder, central hub or a master earth station, and remote VSATs. A VSAT terminal installed at geographically dispersed locations has the capability to receive as well as transmit signals via the satellite to other VSATs in the network. Depending on the access technology used, signals are either sent via the satellite to a central hub, or directly to VSATs, with the hub being used for monitoring and control. A VSAT comprises two units—outdoor and indoor. The outdoor unit consists of an antenna and Radio Frequency Transceiver (RFT). The antenna is typically 1.8 metres or 2.4 metres in diameter, though smaller antennae are also in use. The indoor unit functions as a modem and interfaces with end-user equipment such as standalone PCs, Local Area Networks (LANs), telephones, or Electronic Private Automatic Branch Exchanges (EPABXs). A typical architecture of the VSAT hub network is shown in the following figure.

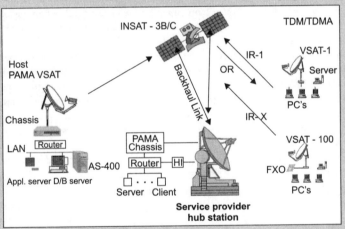

Architecture of the Very Small Aperture Terminal Hub Network

Following are the two types of VSAT networks in use:
1. *One-way Systems*: A one-way VSAT system relies on a transmitting station that transmits one or more carriers to the satellite that rebroadcasts the signal over its coverage area. All receive-only VSATs under the satellite footprint can then receive the signal or the user/operator can define groups of VSATs from one-to-all on the network.
2. *Interactive Systems*: Interactive VSAT systems come in two main network topologies—star and mesh. The star topology is based either on a shared access scheme, which is designed to support transaction-processing applications, or on a dedicated link (the satellite being equivalent to a leased line). The mesh topology usually uses links that are set up and torn down

on request, to establish a direct link between the two sites on a demand-assigned basis. Mesh systems were initially designed to support corporate and public network telephony links, but are being increasingly used to serve high data rate services, such as file downloads.

There are two types of access systems in use. They are: Time-designed Multiple Access (TDMA) and Demand-assigned Multiple Access (DAMA). In a TDMA network, all remote VSATs communicate with the service provider's central hub station. The hub station monitors and controls all VSATs in the network and the entire customer traffic is routed through it, using time division multiplexing (that is, each VSAT is designated a time slot to transmit). The TDMA is typically used in a star topology.

On the other hand, in the DAMA network, VSATs are pre-allocated a designated frequency (also referred to as Permanently-assigned Multiple Access [PAMA]). Equivalent of the terrestrial leased-line solutions, DAMA solutions use the satellite resources constantly. Consequently, there is no call-up delay, which makes them most suited for interactive data applications or high-traffic volume. The DAMA solutions are alternatives to leased-line solutions. The call is set up directly between VSAT terminals and hence in a mesh topology. The DAMA solutions normally use a single satellite carrier for carrying a single channel of user traffic, referred to as Single Channel Per Carrier (SCPC). On demand, a pair of available channels is assigned, to enable establishment of calls. Once the call is completed, channels are returned to the pool for being assigned to another call. Since the satellite resource is used only in proportion to the active circuits and their holding times, it is ideally suited for voice traffic and data traffic in batch mode.

The combination of both star and mesh topology can be achieved by having multiple centralized uplink sites interconnected together.

Source: *Voice and Data*, 30 March 2003 and 5 March 2009.

can also be used to provide access to geographically distributed locations of enterprises. On the other hand, VSAT networks if interconnected to other networks such as the PSTN or the Internet can also provide backbone and long-haul network connectivity, especially for telecom and Internet service providers.

Large national telephone companies such as BSNL use VSATs to extend networks into remote regions. About three to five local access lines are terminated into a centrally-located VSAT terminal. This type of installation, referred to as 'thin-route telephony', is suitable to provide telephony services in a cost-effective way for remote hilly regions of the country (Chhibbar, 2000). Telecom networks in Jammu and Kashmir, most of the Northeast, island territories like Andaman and Nicobar, Lakshadweep and Daman and Diu do not have terrestrial connectivity and hence backbone network connectivity is done through VSAT. A VSAT-based voice service is provided across large areas of Africa (referred to as Aquila). Mass production of VSAT terminals for this project reduced per terminal cost and made it a viable proposition to connect different villages of Africa (Blèret *et al.*, 1998).

Very Small Aperture Terminals in Disaster Recovery

The VSAT networks have been installed for disaster recovery management in states such as Orissa which face cyclones and floods. This is to provide alternative communication channels when the terrestrial networks are destroyed and become inoperable. The Maharashtra government also has installed many VSATs towards this purpose (Chhibbar, 2000). Many enterprises use VSAT links for providing backups for their ISDN, leased line, MPLS networks due to the high availability of VSATs.

Banking, Financial Services, and Insurance Segment: Connecting Automated Teller Machines

One of the widely used applications of VSATs is in the banking, finance, and insurance sector for interconnecting Automated Teller Machines (ATMs) to the banking servers. One of the largest installations of VSATs in the country is by the State Bank of India for connecting its ATMs, especially in smaller towns and villages across the country. Though the transaction cost of for a debit card swipe at an ATM is around Rs 6–7 per swipe, which is much higher than doing it through an ISDN line, the high level of reliability of VSAT makes it a better choice, especially for banks (*Voice and Data*, March 2007).

Connection for Lotteries

The other major application is providing network connectivity to lottery operators in different parts of the country. For example, the lottery operator Playwin has a large number of VSATs in place to connect lottery terminals in different states of the country. While Gilat and HCL Comnet dominate in the ATM connectivity solutions, Hughes Escorts Communications Ltd. (HECL) won major contracts, especially from Playwin for VSAT connectivity in the lottery sector.

Enterprise Connectivity

Oil companies such as Indian Oil, Hindustan Petroleum and Bharat Petroleum use VSAT in petrol pumps located in smaller towns and villages to collect data for quality monitoring. Though the low cost and improved quality of VPNs have reduced the demand for VSAT in enterprise applications, the FMCG companies such as Hindustan Unilever still use VSATs in remote parts of the country for providing access to enterprise applications such as Enterprise Resource Planning (ERP) systems.

Distance Education

Another sector that is catching up is education where institutes such as the Indian Institute of Management (IIM) Kozhikode, IIM Bangalore, Xavier Labour Research Institute, Manipal University and Career Launcher use VSATs for distance education. Hughes Escorts Communications Ltd. is the favoured service provider in this space. With the Ku-band opened up on India's EduSat satellite, the coming days would see more deployment of VSAT hubs for educational purposes. The major user in the education segment is the Indian Space Research Organization (ISRO) which is deploying more VSATs connected through the EduSat satellite transponder for connectivity to more than 50,000 schools in the country. The HealthSat project conceptualized by the Indian government along the lines of EduSat will also use VSATs for telemedicine applications (*Voice and Data*, 13 June 2005).

Rural Community Service Centres

The VSAT technology is also used extensively in the government's Community Service Centre (CSC) scheme in about 100,000 villages in the country. The CSCs are meant to provide people in rural villages access to various government services such as land records and bill payments at kiosks connected via VSATs (*Voice and Data*, March 2009). A significant number of VSATs has been deployed for the e-Gram connectivity project initiated by the Government of Gujarat. The Gujarat government is connecting 13,716 gram panchayats and CSCs as a part of its e-Gram connectivity infrastructure project using VSATs as the last-mile option. The project will ensure high-quality and cost-effective video, voice and data services in the areas of agriculture, e-governance, health, education, and so on. The connectivity will also facilitate point to point and point to multi-point video conference services, VoIP services and intra and Internet services from these village panchayats and CSCs.

Very Small Aperture Terminals in Entertainment

Bollywood producers are already distributing cinema in digital format. In digital cinema, instead of celluloid prints, the full movie is transferred to Compact Discs (CDs) or a hard disk and taken to the theatres individually. However, with satellites and VSAT hubs, the movie can be downloaded at the VSAT network operations centre. Then, on the desired day and time, it can be beamed to theatres simultaneously on a 10 Mbps connection. It will take six to eight hours to download a three-hour movie. After downloading it, the theatre can screen it. The Digital Content Delivery (DCD) service enables the movie producers to transmit movie files as large as 40–60 GB across multiple theatres in a single satellite transmission. This makes DCD a more cost-effective means of distributing movies to theatres across the country as opposed to the traditional forms of delivery. The advantage of this process is that the movie reaches even remote areas on the same day and reduces chances of piracy. Early in 2004, the Bobby Deol-starrer *Bardasht* was transmitted from Mumbai to the Wave Cineplex in Noida, Uttar Pradesh via satellite as a pilot run. Though the pilot was a success, widespread commercial deployment is yet to take place. Though the cost per theatre gets as low as Rs 7,000 per theatre per week (instead of the traditional Rs 60–90,000), the cost of digital projectors and VSAT hardware is high enough—Rs 10 lakh onwards—to keep the theatres disinterested (*Voice and Data*, 6 January 2005).

Very Small Aperture Terminal Market

The VSAT industry in India is highly regulated and hence has not grown in line with the expectations and demand. Competition from Internet-based VPN services, lowering of leased-line tariffs due to high competition in the NLD sector, regulatory restriction on interconnectivity and lack of adequate transponder space are reasons for the slow growth of the VSAT sector in our country.

The growth did not happen until 2004 (see Figure 7.1), thanks to intense competition provided by terrestrial connectivity options such as leased lines and VPNs (for VPNs refer to Chapter 6). The price of leased lines dropped due to the competition in the NLD sector. Even in traditional markets such as manufacturing VSAT was relegated to the second position. More than 1,000 VSATs were surrendered during the year. Also, the much-hyped segments such as rural telephony and broadband Internet access failed to take off (*Voice and Data*, 30 March 2003). However, the market started improving in 2004, due to increase in the availability of Ku-band transponders including those from foreign satellites, increase of speed from 64 Kbps to 512 Kbps and the operators migrating to a revenue-sharing scheme.

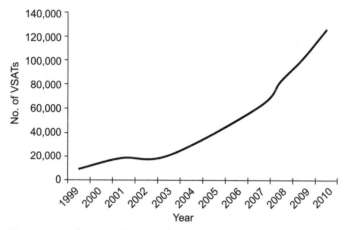

Figure 7.1   Growth of the Very Small Aperture Terminal Market

The market share of the different VSAT operators is shown in Table 7.1. Banking and financial sector (BFSI), retail, small and medium businesses and the recent installations of the rural CSCs are the different market segments to which the VSAT players are catering.

Very Small Aperture Terminal Regulation in India

Licensing and Scope of Operation

The NTP 1999 envisaged grant of licence on non-exclusive basis to VSAT service providers for a period of twenty years extendable one time by ten years. Accordingly, licenses are issued to applicants to establish, install, operate and maintain VSAT Closed User Group (CUG) domestic data network service via the INSAT system on non-exclusive basis within the territorial boundary of India.

There are two types of CUG VSAT licenses: (i) Commercial CUG VSAT licence and (ii) Captive CUG VSAT licence. The commercial VSAT service provider can offer the service on a commercial basis to the subscribers by setting up a number of CUGs whereas in the captive VSAT service only one CUG can be set up for the captive use by the licensee.

The scope of the service is to provide data connectivity between various sites scattered within the territorial boundary of India via the INSAT system using VSAT. However, these sites should form part of

Table 7.1   Market Share of Very Small Aperture Terminal Operators

VSAT Operator	Installed Base (March 2010)	Market Share	Market Segment
Hughes	36,028	29.10%	BFSI, Retail
Bharti	38,380	31.00%	Rural CSCs
HCL Comnet	27,733	22.40%	BFSI
Tata Net	12,628	10.20%	SMB
Essal Shyam	2,724	2.20%	Enterprise, Entertainment
Others	990	0.80%	Captive
Total	123,807		

a CUG. A CUG is permitted to have business association with the following categories (DoT, 2007):

1. Producer of goods and his trader/agent;
2. Provider of service and his trader/agent;
3. Producer of same category of goods (for example, manufacturers of petroleum products); and
4. Provider of the same category of service (for example, bank).

Provided that the ultimate consumer of a service or a product shall not be a part of the CUG; and provided further that the CUG is meant for the legitimate business communication purposes of the group.

The entry fee for VSAT licence is Rs 3,000,000. The annual licence fee for commercial VSAT services is 6 per cent of the AGR. This includes the 5 per cent towards the USO fund. This is in line with the annual revenue fee charged for the NLD and ILD licence holders. For captive VSAT services, the annual fee is Rs 10,000 per VSAT multiplied by the number of VSATs (including transmit-only VSATs, receive-only VSATs and both transmit/receive-VSATs). A minimum of five VSATs along with Hub must be commissioned by the commercial VSAT licensee within a period of one year from the date of grant of licence. In the event of failure, an amount of Rs 1,000,000 shall be recovered for each year's delay up to a period of two years whereafter the licence shall be terminated. Similar penalties apply for non-commissioning of service during the first year of licence. It must be noted that VSAT service providers paid Rs 50,000 per terminal as licence fee until April 2001. The new VSAT policy brought VSAT services under revenue-sharing scheme as practiced in the other telecom services (*Voice and Data*, July 2004). The per VSAT fee regime just about killed the VSAT industry as the growth of the sector was penalized due to the linear fee structure. The new revenue-sharing policy though announced in 2001, was finally practiced only 2004 onwards which reduced the regulatory levy burden on VSAT service providers.

Until 2001, only 64 Kbps data speed was allowed through the VSAT networks. There were two reasons for this myopic policy:

1. There was not enough transponder space in the INSAT system to allow higher bandwidth to be leased by the operators for high-speed data transmission.
2. The VSAT service was not expected to be a substitute for landline-based leased-line services offered at that time, mainly by the government department (Department of Telecom Services, DTS) and government-owned MTNL.

This restriction in bandwidth limited the type of applications to be provided using the VSAT network. Except for transaction-oriented text-based services, none of the emerging web-based services could be provided by the VSAT service providers. In 2001, the new policy hiked the maximum transmission speed up to 512 Kbps per VSAT (including all carriers) in star configuration and up to 2 Mbps in mesh configurations (*Voice and Data*, July 2001). This gave a fillip to the VSAT industry as new and emerging services could now be provided, especially to enterprise clients. This certainly shows in the increasing in number of VSATs deployed and the revenue of the industry 2004 onwards.

Frequency Bands

The VSATs are used for both one-way and two-way interactive systems. The VSAT operations were allowed in India only on the INSAT series of satellite on the extended C-band (5.850–6.425 GHz transmit 3.625–4.200 GHz receive). The C and extended C-band suffer less from rain attenuation, but required larger antennae, and are used in Asia, Africa, and Latin America which are located in the tropical region. However, most of the developed countries in North America and Europe migrated to the use of Ku-band operating in 11–14 GHz, which can use smaller antennae, but suffers from rain fade in a monsoon-like downpour. Typically, the size of an interactive Ku-band antenna ranges from 75 cm to 1.8 metres and that of the C-band ranges from 1.8 metres to

2.4 metres. The lower frequency C-bands have lesser bandwidth of about 36 MHz, whereas Ku-band has a bandwidth of 36/54/72 or 77 MHz. Reduced interference with the terrestrial radio system reduces constraints on the location of the dish antennas to be deployed.

The Ku-band was not allowed for use in India for satellite connectivity until 2001 as there were no Ku-band transceivers deployed on the INSAT systems until then. Moreover, there has been reluctance to use Ku-band in tropical regions like India due to rain attenuation. However, this problem could be overcome by providing adequate fade margin to take care of attenuation during the monsoons. The cost of Ku-band components has also decreased due to large-scale production economies.

In 2000, INSAT 3B was launched which had three Ku-band transponders to start with. The satellite communications policy redrafted in 2000, allowed private ISPs to use Ku-band for setting up international gateways. At the same time DTH guidelines were announced by the Government of India which allowed Ku-band transceivers to be installed at customer premise. Finally, in 2001, the new policy allowed VSAT service providers to use foreign satellite transponders under certain restrictive clauses.

The VSAT guidelines strictly mentioned the use of the INSAT system transponders for connectivity. Table 7.2 gives the number of transponders in the INSAT system.

Table 7.2   Indian National Satellite System Capacity

INSAT	Transponders	Year of Launch
2E	12-C; 5-xC	April 1999
3C	24-C;6-xC	January 2000
3B	12-xC; 3-Ku	March 2000
3A	12-C; 6-xC; 6-Ku	April 2003
3E	24-C; 12-xC	September 2003
4A	12-C; 12-Ku	December 2005
4B	12-C; 12-Ku	March 2007
4CR	12-Ku	September 2007
4E	5-C × S; 5-S × C	Scheduled 2011
4F	C, S and Ku band	Scheduled 2011
4G	18-Ku	Scheduled 2011

Source: ISRO, Gunter.

As can be seen from Table 7.2, the Ku-band transponder capacity is still not adequate to meet the demand of the DTH and VSAT services (for DTH refer to Chapter 8). Realizing this the government allowed the use of foreign satellites for VSAT services. However, there are many restrictions and procedures to be followed in case of the usage of foreign satellites, which are outlined below.

Rules for the Use of Foreign Satellites

Any foreign satellite company can provide bandwidth capacity services to Indian end-users (such as DTH and VSAT service providers) through the ISRO. For this purpose, the transponder capacity is channellized through ISRO and ISRO's approval may be forthcoming if alternate capacity is not available on INSAT systems. As the leasing process involves the DoS, the WPC Wing of the DoT and the concerned Ministry (for instance in the case of VSAT, the Ministry of Communications and Information Technology), the process is definitely time-consuming and with no clear assurance of achieving approval. The foreign satellite should be coordinated with INSAT systems in terms of the ITU's radio regulations as a pre-condition for lease of capacity. A short-term lease (usually on a yearly basis) is signed by the ISRO. There can be no private contract between the end-customer and the satellite company though typically the Indian customer and the satellite company agree on the commercial terms prior to approaching the ISRO. The INSAT reserves the option of not extending the capacity lease for subsequent years and requires the end-customer to migrate to an INSAT system, as and when alternate capacity is available. This indirect route of providing capacity to the Indian market has been adopted by several foreign satellite operators in the past including SES Americom, Asiasat, and New Skies.

Because of the delays and procedures involved in the use of foreign satellite transponder space, most of the VSAT operators are constrained to use INSAT transponder capacity. Lack of supply with an increasing demand for transponder capacity resulted in high cost of services and made the VSAT service unattractive for corporate customers.

Cost of Very Small Aperture Terminal Services

As of March 2009, the approximate tariff for 64-Kbps VSAT connections ranged from Rs 25,000–35,000 (*Voice and Data*, March 2009). The VSAT components are also very expensive. The VSAT antennas and indoor equipment costs Rs 150,000 upward, which makes it unaffordable for SMBs. The VSAT providers have come up with innovative rental schemes to attract customers. However, the VSAT service guidelines set a ceiling price for VSAT services as shown in Table 7.3 (DoT, 2007).

The ceiling charges mentioned in Table 7.3 cannot be carefully monitored as more often the VSAT services are provided on a flat rate basis (irrespective of the amount of data transmission). The only critical aspect of price is speed of transmission about which nothing is mentioned in the table for benchmarking purposes.

Interconnection with Other Networks

There are strict regulatory guidelines for interconnecting VSAT commercial and captive CUG networks with other networks. The interconnection guidelines are given below (DoT, 2007):

1. Interconnection with PSTN not permitted.
2. Network of other VSATs—Interconnection shall be permitted through the hub on case to case basis, wherever the CUG nature of the network is not violated.
3. Terrestrial data lines leased by customers of VSATs—Interconnection shall be permitted on case to case basis, wherever the CUG nature of the network is not violated.
4. Terrestrial data lines of a public nature—Interconnection shall be permitted through the hub provided it is connected to a public data network such as the Internet/INET (the packet-switched network of BSNL).
5. Overseas office of the CUG for data transfer purposes: Interconnection shall be permitted on a case to case basis subject to the condition that the connection should be between the hub and the server of the overseas office through a leased line passing through an international gateway which can be monitored for security purposes.

Table 7.3 Ceiling Charges for Very Small Aperture Terminal Services

Item	Ceiling Price in Rs
Registration Fee	10,000/ VSAT
Data Service	3 lakhs/year/VSAT
Data and occasional voice service in CUG	3.5 lakhs/year/VSAT
Satellite Access Charge	10,000/month/VSAT
Volume Charges	50 for every 64 Kbyte segment

6. Wide area network (WAN) Operators: Interconnection shall be permitted on a case to case basis, wherever the CUG nature of the network is not violated.
7. Internet: The hub of the VSAT licensee shall be allowed to be connected to an Internet node of its choice through a lease line taken from a telecom service provider who is authorized to sell bandwidth/ leased line.
8. Other media to provide for redundancy: Switchover between a terrestrial CUG network and a VSAT-based CUG network belonging to the same licensee shall be permitted for redundancy purpose.

As can be clearly seen, the interconnection between VSAT networks and other networks are provided only when the CUG norms are not violated. Internet connectivity was not allowed until 2001. Due to pressure from the VSAT service providers, the VSAT guidelines were amended in 2001 to allow Internet connectivity. However, interconnection with PSTN is still not allowed and hence no modifications. Interconnectivity with PSTN will allow long-haul voice traffic to be carried across the VSAT network, thus bypassing the traditional NLD carrier network. The NLD carriers have been opposing the VSAT–PSTN interconnectivity due to the high fee of about Rs 100 crores they paid in 2001 as entry fee for their NLD service licence. Compared to that the VSAT operators paid an entry fee of Rs 0.3 crores. However, after the revision of the NLD licence fees in 2005, the NLD carriers pay Rs 2.5 crores as entry fee which is not very high compared to the VSAT entry fee. Hence there is a case for VSAT–PSTN interconnectivity for providing cheaper long-distance voice service, especially in the remote and rural areas of the country.

* * *

For a large country such as India, satellite services hold promise for connectivity, especially in the rural and remote parts of the country. In the next chapter, the role of satellites in the provisioning of broadcasting is discussed.

8

Broadcasting Services
Moving Towards Digitization

Introduction

Cable TV came into existence in India in 1983 when Doordarshan (the broadcast arm of the government) started its services on cable networks in the rural villages of Rajasthan. In 1989 a few entrepreneurs set up small cable TV networks and started local video channels showing movies and music videos. The international satellite television was introduced in India during 1991 with the coverage of the Gulf war by CNN. The spread of cable TV received a boost during 1992 with the launch of the cable TV programme networks by Zee Telefilms and the STAR group by beaming India-specific content. Table 8.1 gives the timeline of various policy and regulatory decisions in this sector.

Market Structure of Cable Television Services

In India, cable TV has adopted a franchisee model with the Local Cable Operator (LCO) being the main contact for subscribers. The LCO lays down the last-mile connection, thus connecting each household to the cable TV network. The LCOs receive the programmes broadcast through the GEO satellites by installing dish antennas at the cable head-end. With increasing number of broadcasters, the number of such antennas also increased and most of the LCOs were not able to install the required number of antennas to provide comprehensive content to subscribers. Hence a set of aggregators referred to as Multi Service Operators

Table 8.1   Sequence of Events in the Broadcasting Sector

Year	Event
1982	First INSAT satellite launched
1983	Doordarshan starts cable TV service
1991	International satellite TV launched
1992	Cable TV programmes launched by Zee and Star
1995	Cable TV Networks Regulation Act announced
1997	Prasar Bharati formed; All India Radio and Doordarshan were brought under it
2000	Phase I of Frequency Modulation (FM) broadcasting; licenses given by auction
2001	DTH guidelines announced by the DoT
2003	First notification of the Conditional Access System (CAS) in Chennai
2004	Broadcasting regulation brought under the TRAI
2004	The TRAI submits recommendations on the second phase of FM broadcasting
July 2005	Phase II of FM broadcasting guidelines announced; auction held
2007	Second notification of CAS in Delhi, Mumbai, and Kolkata
2007	The TRAI releases recommendation on Head-end in the Sky (HITS)
2009	The DoT releases HITS guidelines

(MSOs) emerged. The MSOs aggregate the content obtained from the broadcasters and some times multiplex the local video channels including movies and songs, and then feed the signal to the LCOs. The MSOs are often owned by relatively large business houses which have an interest in the broadcasting business. Examples include Sun Cable and Hinduja's InCable. The distribution chain of the cable TV industry is shown in Figure 8.1.

Market Size

In India, out of the 255 million households, 128 million had TV and 78 million received cable TV services at the end of December 2007 (TRAI, July 2008). The TV industry market size is estimated to be about Rs 23,000 crores. The number of cable TV subscribers is more than the number of landline subscribers in the country, indicating a phenomenal growth, thanks mainly due to lesser regulation and the franchise model of the LCOs. As per the Ministry of Information and Broadcasting, there are 30,000 registered cable and satellite operators in the country,

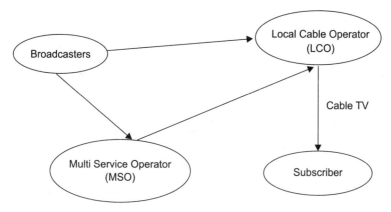

Figure 8.1   Distribution Chain of Cable Television

broadcasting over 339 cable and satellite TV channels in national and regional languages (TRAI, July 2008). Taking into account the number of unregistered cable TV operators, an LCO in India connects about 1,200 subscribers.

The cable TV market has been traditionally considered as a 'natural monopoly' market in much the same way as the fixed wire-line telephony market. Since the coaxial cable laid by the LCO provides the last-mile access to the subscriber, the characteristics of the network in terms of sunk cost, and substitutability are much like the wire-line network. Though there are no restrictions on the cable TV licenses in any geographical area, there is only one LCO in operation in most geographical regions of the country.

Technology of Cable Television

The architecture of a typical cable TV network is shown in Figure 8.2. The LCO installs and maintains the local cable network and gives connectivity to subscriber homes. Since the average number of households connected to the cable TV network is much more in India compared to rest of the world, often devices such as 'one-way amplifiers' are installed, especially in the later legs of the cable TV network to boost the signal strengths to the desired levels.

Conditional Access System

The distribution chain as shown in Figure 8.1 has inherent weaknesses. Since the broadcasters and MSOs do not come into direct contact

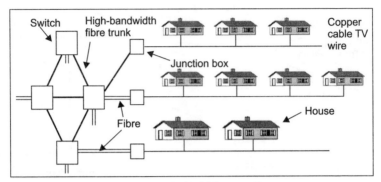

Figure 8.2 Architecture of the Cable Television System
Source: Tanenbaum, 2001.

with the subscribers, there is information asymmetry between them and the LCOs who is the direct contact with the subscribers. This information asymmetry was exploited to a certain extent and the LCO could potentially hide the exact number of households connected, thus saving the revenue to be shared with the broadcasters and MSOs. The MSOs in turn might disconnect signals to the LCO without any prior notice and seek undue enhanced commitment for subscriber base and higher payments. Moreover, LCOs are required to pay entrainment tax and service tax which are linked to the number of subscribers. Without the implementation of proper billing systems, LCOs may evade taxes by under-declaring their subscriber base (TRAI, July 2008).

Realizing these problems, the Conditional Access System (CAS) was proposed to be implemented in 2003. A Notification dated 14 January 2003 was issued by the Ministry of Information and Broadcasting, Government of India, making it obligatory for every cable operator to transmit/re-transmit programmes of every pay channel through an addressable system in the Chennai Metropolitan area, Municipal Council of Greater Mumbai area, Kolkata Metropolitan area and the National Capital Territory of Delhi within six months from 15 January 2003 (TRAI, January 2004). However, due to political haggling and the dispute between LCOs, broadcasters and the MSOs, CAS was rolled out only in the Chennai metro area and withdrawn in other areas of the country. Subsequently, the government pressed the CAS service in select areas of Delhi, Kolkata, and Mumbai in January 2007. After the initial phase, the CAS was expected to be rolled out in other areas of the country in subsequent phases.

The CAS introduced addressability in the cable TV network along with providing signals in digital format. The CAS ensures that only duly authorized subscribers are able to view a particular programming package. A CAS system consists of an integrated receiver decoder also called the Set-Top Box (STB) at the subscriber premise. This is an electronic device which contains the necessary hardware, software, and interfaces to select, receive, unscramble and view programmes. Since signals are scrambled in CAS, only the viewers with a valid signed contract with CAS service providers are authorized to unscramble and view the chosen programmes. Moreover, when the viewer chooses a pay channel or a programme, the information related to subscriber details, method of payment, and services purchased is stored and updated in the database. Apart from selecting pay channels, a CAS may be used for provisioning of other services such as video on demand.

One of the intended benefits of the CAS is that the consumer is able to choose channels of his/her choice. He/she can either opt for the package of free to air (FTA) channels or take his/her choice of pay channels.

The history of the CAS rollout, and the politics and economics of the CAS is explained Box 8.1 (Sridhar and Sridhar, January 2008).

Regulation of Cable Television

Post CAS there were many problems:

1. The cable TV industry evolved to the current structure in an unorganized manner. Though the phenomenal growth of the industry is attributed to the entrepreneurial skills of the LCOs, it brought in its wake certain consumer-related problems.
2. Despite the thousands of LCOs operating in the country, there is no competition in the last mile and hence no choice for the consumer in cable TV. This has given rise to increasing bills and poor customer service in most locations of the country.
3. Post CAS, there are problems with respect to customers having to take the entire bouquet of channels made available by the LCO or nothing at all. Price of pay channels increased leading to customer distress. The STBs were not interoperable making it difficult for the subscribers to reuse the STBs when shifting to different regions of the country.
4. Though an alternative to cable TV such as DTH was allowed, it did not take off due to high prices.
5. Cable TV penetration in rural and remote areas of the country was still poor.

Box 8.1   The Politics and Economics of the Conditional Access System

The much controversial CAS that was supposed to be introduced in the country way back on 14 July 2003 has still not seen the light of day. Even in its limited implementation in Chennai and South Delhi it has raked up more controversies and confusions than any other event in the history of the Indian TV industry. Meanwhile, the Government of India issued a notification amending the TRAI Act and authorizing the TRAI to look into all issues relating to the CAS. We address some of the issues raised in the consultation paper released by the TRAI.

First, why CAS? As in the classic principal-agent problem, the under-reporting of subscriber base by LCOs was the motivation for broadcasters to initiate CAS so that they would be able to get their appropriate share of the subscription revenue. The government too was interested in moving towards CAS to stop leakage in the service tax paid by the cable operators.

Second, through CAS, strong consumer preferences for certain programmes can be tapped, and such consumers can be made to pay a premium price for their preferred programmes. The desirability of CAS also stems from the free-rider problem, and the tendency of many consumers to view their most preferred channels at minimal charges. Stronger preferences will lead to adoption of CAS while weaker preferences will persuade the consumers to stay with FTA channels that can be viewed without the STB. Direct-to-home when introduced in the US in the early 1990s was a big hit as households had the luxury of receiving the channels they wanted and were relieved to be out of the clutches of monopoly cable operators!

What are the essential characteristics of the consumers who opt for CAS? These are mostly households with the ability and willingness to pay for the channels they prefer to view. Demographic profile is certainly the other factor. Note that consumers with special preferences such as those with flexible jobs, housewives (soap operas on Zee and Sony), and children (Cartoon Network and Nickelodeon) are the ones who are likely to adopt CAS for their preferred channels. Finally, we observe that CAS is likely to be successful only for consumers educated beyond a certain level. For instance, consumers who are not very well-informed, are merely 'couch potatoes' (TV addicts) for whatever programme that appears, and those who do not understand the implications, would just prefer not to pay for CAS, as long as they have a functioning TV! So, CAS is ideal for rollout only in metros, and cities and not obviously in rural areas. The FTA channels distributed in the CAS areas should be mandatorily broadcast in non-CAS areas also. In non-CAS areas, the cable operators can judiciously mix other channels to suit the interests of the local population.

From the consumer's point of view, CAS can be considered as a normal good. Its income elasticity of demand is positive. Metros and urban areas with increased disposable income among the populace, will have higher probability of subscription to CAS. Price elasticity of demand for CAS is negative, which implies that even CAS-preferring consumers would not want all pay channels, and would likely back out if the only option were to purchase all pay channels, and this price were to be set high. Hence regulation should force broadcasters to offer a la carte channels and prescribe a ceiling price as a percentage of the bouquet so that the consumers' interests are protected and that they are not taken for ride by the broadcasters. The ceiling price for non-CAS areas also needs to be specified by the regulator as cable operation still is far from competition for the market to determine the equilibrium price.

The other determining factor for the success of the CAS is the optimal mix of advertising and subscription revenue the broadcaster gets. Out of the TV industry revenue of Rs 111 billion in 2002–3, Rs 60 billion was from the subscription revenue. While broadcasters of FTA channels earn their revenue through advertising, pay channel broadcasters get both subscription and advertising revenues. However, if CAS subscription does not pick up, even the advertisers will turn their back on the pay channels and move over to FTA channels for viewer base. So broadcasters have to tune in to the needs of the customers more carefully than ever before. They have to learn to do better yield management by choosing optimal pricing strategies for advertising and subscription. Hence the success of CAS depends on the price of pay channels and the ability of the broadcaster to provide niche content. Fixing upper limits for advertising time for FTA channels might force the broadcasters to move their channels from FTA to pay. Hence the regulator should specify the upper limits of advertising time only for the pay channels, to protect the viewing interests of CAS consumers.

The CAS should bring not only the preferred channels to homes but also the associated quality of service. The regulator should specify strict norms of quality for pay channels and appropriate penalty clauses for transmission impairment so that the cable operators become more accountable.

Finally, is CAS needed? It is very similar to users paying additional charges for certain value-added services such as for SMS in cellular service. The CAS makes broadcasters and cable operators accountable for their services which was not the case before. Hence the government should break the imbroglio and go for its full implementation. If it rolls back CAS in south Delhi, due to its populist attitude as it is with most other policies, confusion will continue to prevail regarding the government's stand, as with many policies. This is quite similar to constructing a multi-lane highway, and not persuading road-users enough, to pay toll. When can we ever expect the reforms to percolate to consumers?

Considering all the above issues, Broadcasting and Cable Services were brought under the purview of the TRAI with effect from 9 January 2004, after the appropriate amendment of the TRAI Act, 1997. This amendment added a proviso to include Broadcasting Services and Cable Services within the scope of 'Telecommunication Services'. Further, the Government of India also issued an order, which mandated the TRAI to make recommendations regarding the terms and conditions on which the 'Addressable Systems' shall be provided to customers. The order also provided powers to the TRAI to specify standard norms for, and periodicity of revision of rates of pay channels, including interim measures (TRAI, August 2004).

Local Cable Operator Licensing Issues

The cable TV industry was provided legal framework through the promulgation of the Cable Television Networks (Regulation) Ordinance, 1994 (which was later converted into the Cable Television Networks (Regulation) Act, 1995). As per the Act, an LCO is required to register with the Head Post Master of the Head Post Office of the concerned area for giving services to the subscribers. The process of registration is simple and the scrutiny is restricted to the submission of certain basic information with corroborative documents.

The area of operation of the LCO is not clearly defined. The LCO may have to seek multiple registrations from the post offices in order to provide services in the entire service area of a district. Duration of registration for the LCO is only one year, with a provision of renewal. The registration fee for the LCO is Rs 500. The fee was set in 1995 and has not been changed as yet.

The LCOs lay cables in their area of operation for provisioning of service to their subscribers. The cables are hung over rooftops and telephone poles. The LCOs have no alternatives as they are not entitled to seek RoW in many states. It is sometimes argued that as LCOs are not licensed under Section 4 of the Indian Telegraph Act 1885, they are not eligible for seeking RoW.

Multi System Operators' Licensing Issues

With an ever-increasing number of broadcasters, the LCOs were required to install multiple dish antennas, which would necessitate huge space and investment. This resulted in the entry of the MSOs who established their head-ends for receiving signals from broadcasters and distributing them to the LCOs. The LCOs either became franchisees or had commercial agreements with the MSOs.

Presently, MSOs are registered as cable TV operators though their operations and business model is different. The MSO's main function is to aggregate content from different broadcasters and provide it to the LCOs. However, as per the existing definition in the Cable Television (Regulation) Act 1995, the MSO means a cable operator who receives a programming service from a broadcaster or his authorized agencies and re-transmits the same or transmits his own programming service for simultaneous reception either by multiple subscribers directly or through one or more cable operators. This means that the MSOs can even directly re-transmit their programmes to the subscribers much like the LCOs. This created confusion over the roles of the MSOs.

The MSO can provide signals within the area of jurisdiction of the Head Post Office in which it is registered. There are several multi-city MSOs which require registration in multiple post offices. Looking at the functionality and business models of the MSOs, a sufficiently large service area is required to attract financially and technically sound entities. The large areas will also provide the economies of scale required to build digital transmission systems and increase the availability of MSO signals in the rural and remote areas of the country. There is no restriction on the number of MSOs in a service area.

There is no entry or annual fee for the MSOs apart from the nominal registration fee payable at the time of registration and renewal. Since the MSOs have to make a substantial investment for setting up their head-ends and to spread their networks, it may be necessary to fix a relatively large entry fee so that non-serious players do not enter into this service space.

Restructuring of Cable Television Services

The above indicates the lack of a robust regulatory framework and licensing conditions for both the LCO and MSO services in the country. Though lack of regulation is cited as one of the main reasons for the proliferation of more than 6,000 MSOs and 60,000 LCOs, the cable TV industry at this stage is mature enough to warrant some regulatory framework for its sustainable growth.

Realizing this, the TRAI initiated a consultation process in early 2008 for restructuring the cable TV services in the country to bring in transparency and accountability. The regulator also initiated steps to improve the quality of service for the consumers in the hitherto unregulated industry segment. The important features of these recommendations are given in Table 8.2.

Table 8.2   Recommendations on Restructuring of Cable Television Services

S. No.	Existing (as of March 2009)	Recommendations by the TRAI (July 2008)
LCO Operations		
1.	Registration for LCO is valid for one year on a renewable basis done at the Head Post Office at the concerned area. Multiple registrations at different post offices required for servicing the district.	The Senior Superintendent of Post Offices in whose area the revenue district falls shall grant District-level licenses. For State-level licence, Chief Post Master General of the Circle shall grant the licence.
2.	The service area of the LCO is not clearly defined.	The service area for cable TV service licence should be the revenue district or the geographical boundaries of the State as the case may be in accordance with the licence issued.
3.	Registration fee for the LCO is Rs 500 per year, payable on renewal.	Entry fee for Cable TV service is Rs 10,000 for District-level licence; Rs 100,000 for State-level licence. An administrative cess of 10 per cent of the entry fee would be levied by the licensing authority for administrative expenses. No annual licence fee.
4.	No quality of service regulation.	The cable TV licensees shall abide by Section 9 of the Cable TV Network (Regulation) Act 1995, in the use of standard equipment in their networks. They are also required to comply with BIS 13420 Part I relating to system performance.
5.	No RoW permission for laying down their local cable network.	The LCOs are eligible for laying copper cable/optic fibre both underground as well as over poles on a non-exclusive basis. Corresponding local bodies/state governments may consider formulating a RoW policy for the LCOs.
MSO Operations		
1.	The MSO is registered as another LCO. Hence there is confusion over whether the MSO can provide signals directly to subscribers.	An MSO means any person who manages and operates a multi-system cable television network to provide cable television service to one or multiple local cable TV operators or to any other distribution platform permitted and licensed by the government. Hence a separate licence is to be issued to the MSOs.

2. Service area is limited to the jurisdiction of the Head Post Office in which it is registered.	The licence area of the MSO can be the area of the revenue district, geographical boundaries of the State or the entire country as a whole, as the case may be.
3. No entry or annual fee.	Entry fee for district-level MSO shall be Rs 100,000; for state-level Rs 1,000,000; and for country-level Rs 2,500,000. However, there is no annual fee for the MSO.
4. No obligation on digitization.	The new MSOs were mandated to digitize their cable TV network within three years from the date of issue of licence as part of their performance obligations, or within five years from the date of notification of the new licensing regime, whichever is earlier. Existing MSOs be mandated to digitize their cable TV network within five years from the date of notification of the new licensing regime.
5. No RoW permission for laying down their cable network.	The MSOs shall be made eligible for seeking RoW on a non-exclusive basis for laying optic fibre.

Source: TRAI, July 2008.

Digitization of Cable Television

The cable TV system in most parts of the country is still analogue. Though CAS in notified areas brought in digitization of networks to a large extent, the penetration was limited to just the four metros as of June 2009. Digital TV transmission offers a number of advantages over analogue. These include better reception quality, increased channel-carrying capacity, new features such as programming guides, multi-view and interactive services, and triple-play services such as voice, data, and video. Most of the television content is produced using digital technologies. Further distribution of content through DTH and IPTV is digital. An example of broadband and Internet connectivity through existing cable TV network is given in Figure 8.3.

Since cable TV covers more than 70 million households, it is imperative to digitize cable TV networks both for leveraging on the technology advantage and for this industry to compete effectively with other digital transmission modes. With this in mind the TRAI released the recommendations for digitization of cable TV networks, way back in September 2005 (TRAI, September 2005). The objective was to promote active digitization of the existing analogue network before the Commonwealth

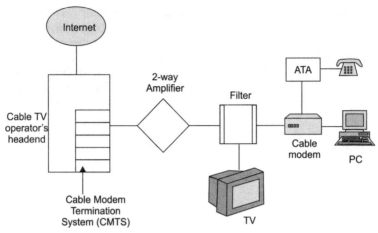

Figure 8.3   Broadband and Internet Connectivity through Cable Television Network
Source: TRAI, March 2008.

Games held in Delhi in 2010. However, as there was no mandate given to the MSOs and LCOs to migrate their networks to digitization, the process is yet to commence in most parts of the country. The TRAI recommendation also clarified the Head-end in the Sky (HITS) guidelines so that HITS can be promoted as one of the alternatives to large-scale digitization of the cable networks in India. Details about HITS are presented in the subsequent section.

However, it must be noted that even the US switched over to fully digital transmission only recently (on 12 June 2009). While the US forced all the broadcasters and TV transmission stations to switch to digital, the TV transition applied only to full-power broadcast TV stations—stations that use the public airwaves to transmit their programming to viewers through a broadcast antenna (DTV, 2009). The cable TV operators were not required to upgrade their cable head-end to transmit digitally. The main objective behind this drive by the US was to free up radio frequency spectrum originally used for analogue transmission for commercial services including mobile emergency calling.

In India, though cable TV connects a large number of households, many viewers, especially in rural areas still use TV antennas to receive over-the-air TV signals. While digitization of cable TV network provides solutions for more than 70 million cable TV homes, the regulator and policymakers need to think along the lines of US regulators to make sure all the broadcasts are in digital form so that they reach the millions of homes who are yet to be connected to the cable TV network.

Direct-to-home Service

Satellite TV in India started way back in the 1960s, when Vikram Sarabhai initiated a national satellite communication group in 1968. Based on its studies and recommendations, the Government of India in 1969 approved a proposal for the Satellite Instructional Television Experiment (SITE) using the National Aeronautics and Space Administration's (NASA's) Application Technology Satellite–6. The main objective of SITE was to provide education to the rural areas of the country. Subsequently, INSATs were launched, starting in 1982, for TV and radio broadcasting. By 1985, 60 million Indian people watched TV programmes and about 50 per cent of the total population was covered by television broadcasts. By 1990, about 90 per cent of the Indian population had access to TV broadcasting due to satellite transmission of TV programming to low-power television transmitters (Contractor *et al.*, 1988).

As described in the chapter on satellite services, satellite transmission was allowed only in C and extended C-band until 2001, when the DTH guidelines were released by the government. The prohibition of the reception of Ku-band was withdrawn by the government on 9 January 2001, allowing for small Ku-band receivers to be installed on home rooftops to receive TV signals directly from the satellite.

The DTH broadcasting service refers to distribution of multi-channel TV programmes in Ku-band by using a satellite system to directly provide TV signals to subscribers' premises without passing through an intermediary such as a cable operator (DoT, 2001). The DTH service cannot be used to transmit signals to cable operators. Hence the business model of DTH does not involve any revenue-sharing with other stakeholders such as MSOs in the value chain, as is often the case with LCOs.

The DTH guidelines took a very protective approach of not allowing more than 20 per cent Foreign Direct Investment (FDI) in the foreign equity of the DTH licensees. An entry fee of Rs 10 crores and an annual levy of 10 per cent of the revenue was prescribed for DTH licensees. The licensee was required to obtain permission from the WPC wing of the DoT and the DoS for clearance and to obtain an operational licence from the WPC before the provisioning of the service. Though the licensee was allowed to use bandwidth from any satellite, preferential treatment was given to licensees who acquired bandwidth from INSAT. However, the TRAI has been insisting that an open-air policy be allowed so that the DTH operators will be able to acquire transponder space from any foreign satellites without much procedural delay (TRAI, January 2008c). Table 8.3 lists the major INSATs that carry Ku-band transponders for DTH services.

Of the above, a glitch in INSAT 4B in 2010 forced shutdown of 50 per cent of the onboard transponders. India launched the INSAT 4C on 10 July 2006. However, the launch was unsuccessful as the Geosynchronous Satellite Launch Vehicle (GSLV-F02) carrying the satellite veered from its projected path and was self-destructed. The INSAT 4CR was launched as a replacement for INSAT 4B.

In the US, although DTH providers first offered services in the 1980s, it was not until the 1990s that they achieved the position of serious competitors to the cable companies. There are two major DTH broadcasters in the US, namely DirecTV (started by Hughes Networks) and EchoStar.

The DTH service though allowed since 2001, the adoption rates both by the service providers and subscribers were very low in India. It was only in 2006 that the subscribers started growing with the entry of new DTH operators. In 2010, there were more than 22 million DTH subscribers distributed amongst five DTH service providers. It is estimated that by 2013, the subscriber base would reach 35 million and hence would possibly overcome the number of DTH subscribers in the US. The list of DTH providers and their subscriber base is given in Table 8.4.

The new MPEG4 standard is being used by most of the DTH operators. The digital video disc (DVD) quality picture and crystal clear sound quality make MPEG4-enabled DTH services a preferred option.

Table 8.3   Indian National Satellites for Direct-to-home Services

INSAT	Transponders	Year of Launch
4A	12-C; 12-Ku	December 2005
4B	12-C; 12-Ku	March 2007
4CR	12-Ku	September 2007
4G	18-Ku	Expected 2011

Table 8.4   Direct-to-home Service Providers and their Subscriber Base
(as of March 2010)

DTH Service Provider	% of Customer Base
Dish TV (Zee)	30
Tata Sky (Tata)	22
Sun Direct (Sun)	25
Big TV (Reliance)	13
Direct TV (Bharti Airtel)	8
D2H (Videocon)	2

Programming Content

It is to be noted that both CAS and DTH are addressable systems and the users pay for the channels they subscribe to. The DTH service is a substitute for CAS as it bypasses the local cable network to deliver TV programmes directly to subscribers. However, it is to be noted that in CAS-notified areas, the LCO has to offer a 'basic service' that consists of about thirty FTA channels and the subscribers should be able to access these basic service offerings without the need for STBs. However, in case of the DTH service, all programmes are viewable only by authorized subscribers through the STB. Even the FTA programmes of the national broadcaster Doordarshan can be viewed by DTH subscribers only through a STB. While it is mandatory for CAS operators to provide channels to subscribers on an a-la-carte basis, it is not so in the case of DTH services (TRAI, March 2009).

The DTH service guidelines stipulate that: 'the licensee shall provide access to various content providers/channels on a non-discriminatory basis'. The non-discriminatory clause does not include 'must carry' conditions. Over the last few years, the number of channels broadcast has increased tremendously. Apart from ground-based channels, there are more than 270 satellite channels registered under the uplinking/downlinking guidelines of the Ministry of Information and Broadcasting and close to a hundred channels are awaiting permission. Since the trans-ponder space for carrying channels is limited, it is not possible to include a 'must carry' clause for the DTH service operators to include all available channels in their offerings. Hence it is clarified in the TRAI's recommendations included 'non-discriminatory' conditions refer to transparent, predictable, fair, equal and unbiased treatment of the different channels by the DTH operator (TRAI, 2008c). This essentially means that the DTH service operator should select the channels for carriage on its platform in a fair and equitable manner, which would enable various content providers to constructively negotiate. The factors that would have a bearing are price and the broad terms offered by the broadcasters. Any decision based on the above-mentioned considerations is further subject to the technical limitation on the number of channels that a DTH platform can carry.

Interoperability

The DTH service licence agreement stipulates that: 'The open-architecture (non-proprietary) set-top box, which will ensure technical compatibility and effective interoperability among different DTH service providers, shall have such specifications as laid down by the Government from time to time'. The requirement of technical interoperability essentially protects the interest of the subscribers by enabling them to shift from one DTH service provider to another without having to buy new hardware. The regulation also requires the DTH service providers to give an option to their subscribers for obtaining the DTH hardware on hire purchase or rent basis. Thus the DTH subscribers have an option to change their service provider through commercial interoperability as provided by the quality of service regulation.

However, provisioning of interoperable STBs has not been successful in India. The main reason for this is the unavailability of Conditional Access Modules (CAMs) of different DTH service providers. The Bureau of Indian Standards (BIS) specifications for DTH STBs require each STB to have a common interface (CI) slot for the purposes of technical interoperability. Technical interoperability is achieved by plugging in the CAM of the new DTH operator in the CI slot of the STB provided by the existing DTH operator. For example, a subscriber of DTH operator 'A' who wishes to switch over to DTH operator 'B' has to procure a CAM from 'B' and plug the CAM into the CI slot of the STB supplied by 'A'. This enables the subscriber to start receiving the services of 'B' using the existing STB and dish antenna (although the dish antenna has to be realigned towards the satellite being used by 'B'). The CAMs presently cost almost as much as a new STB. Therefore, technical interoperability has not been very successful. However, the new entrants are expected to make the CAM available and with the increasing DTH subscriber base, the demand for CAM is expected to increase resulting in a price decline.

These two factors should promote interoperability of STBs amongst DTH operators. It is also ideal that the STBs be interoperable not only amongst DTH operators but also across CAS and IPTV providers. Both the CAS and IPTV markets are at such a nascent stage that this is not likely to happen soon.

The other issue is regarding the MPEG-2 and MPEG-4 digital compression formats used by the DTH operators. The newer MPEG-4 format delivers DVD-quality video at lower data rates and smaller file sizes. While the older operators still use MPEG-2 format, the new entrants have migrated their transmission to MPEG-4 format, thus using superior compression technology to save as much as 25 per cent of transponder bandwidth. This change in formats implies that the subscribers who currently subscribe to MPEG-2 cannot migrate to DTH operators who use MPEG-4 though the converse is possible. The solution to this problem as outlined by the TRAI (2008c) is that the DTH operators should migrate to any new standard set by the BIS within a stipulated time. Hence the existing MPEG-2 STBs shall be changed to MPEG-4 by the DTH operators within the timeframe. However, they can continue to broadcast their content in MPEG-2 until such time as they shift over to the new standard.

DTH and Broadband

One of the most important recommendations put forward by the TRAI was regarding the use of the DTH platform for providing broadband services (TRAI, January 2008). The DTH service is primarily meant for providing downlink broadcast services. However, as invented by Hughes Networks in the US, the DTH platform can be used to provide high-capacity downlink broadband connectivity. The uplink of information to the ISP node or the network interconnection point can be through an alternative access mechanism such as GPRS-based mobile service. The advantage of this approach is the provisioning of high-speed Internet service up to 1.5 Mbps (since most of the Internet applications require high downlink speed compared to uplink). Moreover, it is to be noted that the following does not require any additional infrastructure at the subscriber premise and offers an opportunity to the DTH providers to make this available as a value-added service making it an attractive option. The architecture of the DTH services for providing Internet access is shown in Figure 8.4.

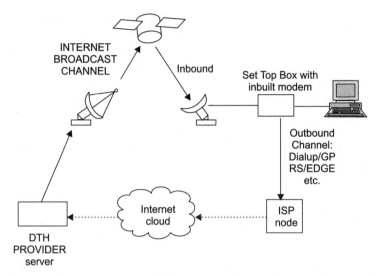

Figure 8.4 Architecture of High-speed Downlink-based Broadband Service through Direct-to-home Service
Source: TRAI, January 2008.

KEY DRIVERS AND CHALLENGES TO DIRECT-TO-HOME SERVICES

The following are the key drivers for DTH services in India:

1. Reach: Extreme geographic reach. Ability to be set up anywhere.
2. Mobility of the STB: The STB and antenna can be used in any part of the country as the subscribers do not have to go through the process of looking for another LCO when they shift residences.
3. Interactive services: A variety of interactive services are offered to the subscribers covering subjects like news, sports, games, entertainment, cooking, astrology, education, stories, and matrimony which when bundled provide a compelling option for the subscribers.

However, there are some challenges as well for DTH services:

1. Satellite transponder space: With regulations regarding the use of satellite transponder space and inadequacy of transponder capacity in the INSAT system, the DTH operators are not able to expand their offerings thus losing their competitive advantage.

Digitization of Cable Systems and Head-end in the Sky

There is an existing analogue cable TV network with about 60,000 LCOs and about 6,000 MSOs, connecting about 71 million homes. As technology progresses and the number of channels downlinked increases in number, it is essential to pave the way for digitization of this cable TV network.

Head-end in the Sky

Head-end in the Sky (HITS) is one of the delivery platforms distributing digital TV content that meets the requirements of digital reception and addressability to cable TV operators. This can be one of the options to spread digitization throughout the country at one go since HITS is satellite-based and hence has a countrywide footprint. The HITS is considered as an important means of digital delivery of TV channels. Hence while the TRAI framed the guidelines for HITS, the basic objective was digitalization of the cable TV network in the country. After consultation the TRAI released recommendations for the implementation of HITS (TRAI, October 2007).

MODELS OF HEAD-END IN THE SKY SERVICES

There can be broadly three models by which services through the HITS platform can be provided. In the first model, the HITS operator contracts with different broadcasters for buying the content, aggregating the same at an earth station and then uplinking with his own encryption to a satellite hired by him. The uplinked channels are then permitted to be downlinked by the cable operators using a dish antenna for onward distribution through the conventional last-mile cable network to homes. In this model the HITS operator works like a conventional MSO, except that virtually the head-end is in the sky, instead of being located on the ground.

In the second model, the HITS operator merely provides passive infrastructure facilities on his satellite to one or more MSOs or to a consortium of cable operators/MSOs desirous of uplinking TV channels to his HITS satellite for downlinking and further transmission to the TV homes by the cable operators across the country. The infrastructure facilities would normally consist of transponder space on satellite, earth station facilities and the provision for mutlicrypting of channels aggregated by different MSOs with different encryption systems. The HITS operator in this model does not contract with the broadcaster for content. He only enters into contract with one or more MSOs or a consortium of cable operators desirous of uplinking their aggregated channels from HITS earth stations to the HITS satellite. In this model, the HITS operator acts as a facilitator by providing the facility of a satellite for the aggregated content to be uplinked and subsequently downlinked by the cable operators.

In the third model, which is a hybrid of the first two models, the HITS operator is free to use his satellite's transponder capacity both for transmitting his own aggregated content, as well as to provide passive infrastructure to other MSOs for uplinking/downlinking their aggregated content.

The TRAI, after the consultation process allowed the third hybrid model whereby the HITS operator can work both as a conventional MSO (except that the head-end is in the sky) as well as passive infrastructure provider to other MSOs/cable operators who wish to use the facility for uplinking/ downlinking their own aggregated content.

Difference between the Head-end in the Sky and Direct-to-home Services

While the HITS is similar to the DTH service in the delivery of channels through the satellite, it is meant to supplement the cable TV network and not act as a substitute as in the case of DTH. It is clearly stated in the recommendations and later in the DoT guidelines (DoT, 2009) that the HITS operator should provide signals directly from the satellite only to the registered MSOs/cable operators. However, the operator will not be barred from providing signals, through its own cable network, if any, to consumers also after first downlinking the signals to the terrestrial station. The HITS operator under no circumstances should provide signals directly from his satellite to the consumer much like DTH service providers. The function of HITS is very similar to the MSOs except that the head-end is terrestrial in the case of the MSOs while it is in the sky in HITS. Both carry and distribute broadcast television signals by first uplinking from an earth station to a satellite in the sky for downlinking at various locations.

Initially it was conceived that band separation for DTH and HITS services was required to prevent HITS operators from providing services directly to subscribers. While DTH operators were restricted to use only Ku-band, HITS licensees were restricted to use C-band. However, the regulator in its later recommendations recognized that band separation between DTH and HITS is not viable in the long term (TRAI, October

2007). Considering the restricted Ku-band transponder space in the INSATs, the TRAI recommended that no band restriction be placed on the HITS as it would delay the digitalization of cable TV networks. However, the HITS operator should supply signals only through the Quadrature Amplitude Modulation (QAM) STBs which are present only at the head-end of the cable operators.

In case of the DTH licence, a one-time non-refundable entry fee of Rs 100 million has to be paid whereas the MSOs as per the new recommendations (TRAI, July 2008), pay Rs 2.5 million for pan-India presence. The investment required for the HITS operation is much higher than that required by the MSOs in providing terrestrial TV broadcasting. The argument is that in order to prevent non-serious players from getting the licence there should be an entry fee. However, because of the nature of operations, it should be less than that of DTH, but higher than that levied for the MSO licence. The DTH operators also pay an annual fee equivalent to 10 per cent of AGR. On the other hand, the MSOs and LCOs do not pay any annual licence fee. One reason for the above asymmetry is that while DTH operators reach their subscribers directly, the MSOs and LCOs have to share their revenues with each other due to the longer supply chain. Hence it is argued that HITS operators who have the same supply chain as that of MSOs should also be not required to pay any annual fee. Considering these various views of the stakeholders concerned, the TRAI in its recommendations (TRAI, July 2008) stated that the entry fee for HITS should be Rs 100 million and that there should not be any annual licence fee to ensure its financial viability and to effectively compete with the terrestrial MSOs.

As in the case of DTH, the HITS operators are allowed to provide any value-added services (for example, bi-directional Video-on-Demand [VoD] and electronic programmable guide [EPG]), which otherwise do not require any specific licence or permission. However, services such as voice, data, Internet which require a separate licence or permission cannot be provided by the HITS operator.

There is no 'must carry' provision for the HITS operator, except for the carriage of some channels of national importance of the public service provider Doordarshan as in the case of DTH operations. However, broadcasters 'must provide' their channels for any distribution platform such as DTH, HITS or the cable network. The TRAI in its recommendations (TRAI, July 2008) noted that in the absence of such a provision, it would be difficult for the distribution platform to source content, and as a result, that platform operator will not take off in the absence of such popular content. Table 8.5 summarizes the different licensing conditions for MSO, DTH, and HITS operations.

Though two teleport licenses have been given for HITS operations, there is not much activity still in the HITS operations in the country.

Frequency Modulation Radio Broadcasting

Radio is still the most popular and affordable means for mass communication, entertainment, and education. The terminal devices are affordable. Even the low-end mobile phones have frequency modulation (FM) tuners and hence the terminal portability and availability is one of the reasons for continued interest in this segment of broadcasting. Though the radio dipped in popularity after the diffusion of TV amongst the households, thanks to the government's initiatives in allowing private parties to enter into this segment it has regained its lost place.

In India, the radio coverage is available in amplitude modulation (AM) (both short wave (SW) and medium wave (MW)), and FM modes. In terms of coverage, MW broadcast covers almost 99 per cent of the Indian population, and FM covers about 40 per cent of the population and 25 per cent of the geographical area of the country (TRAI, 2008a).

As an initial step towards consolidating efforts in public service broadcasting, the Government of India created 'Prasar Bharati', as a

Table 8.5 Categories of Broadcasting Licensees and their Characteristics

Item	MSO	DTH	HITS
Downlink Target	Terrestrial broadcast to head-end of cable operators or directly to subscribers	Satellite downlink only to subscribers	Satellite downlink to the head-end of MSOs or LCOs; can also downlink to its own terrestrial cable head-end for redistribution through the terrestrial cable network; no downlink directly to subscribers
Entry Fee	Rs 2.5 million for countrywide presence	Rs 100 million	Rs 100 million
Annual Fee	Nil	10 per cent of AGR	Nil
Downlink Band	Not Applicable	Ku-band	No band restriction

statutory autonomous body established under the Prasar Bharati Act. The Board came into existence in November 1997. The Prasar Bharati is defined as the public service broadcaster of the country. The objective of public service broadcasting is to be achieved through All India Radio (for public radio) and Doordarshan (for public TV) which earlier were working as independent media units under the Ministry of Information and Broadcasting.

In the year 1999, the Government of India decided to allow private sector participation in FM radio broadcasting. Until then the government entity All India Radio was the sole radio broadcaster in the country. The focus of the Ninth Five-Year (1997–2002) Plan was on improving the variety and quality of radio broadcasting in the country and the focus of the technology shifted from MW to FM due to superior quality of FM broadcasting. In May 2000, the government started the first phase and identified 108 frequencies in the FM spectrum (87–108 MHz) for auction across forty cities in the country. The cities were categorized as given in Table 8.6. Details of the auction are given in Table 8.7.

Though bids were received for 101 frequencies, finally only twenty-one channels spanning just twelve cities were operational. One of the main reasons cited was the high licence fee and hence unviable business models. To analyse the outcome and define the path forward, a committee named the Radio Broadcasting Policy Committee was set up in July 2003. In February 2004, the TRAI was asked by the Ministry of Information and Broadcasting to give guidelines for Phase II of private FM radio licensing. In August 2004, the TRAI submitted its recommendations. Based on these recommendations and the radio policy committee report, the Phase II policy was announced in July 2005, placing for bid, 337 channels encompassing ninety-one cities. After scrutiny, finally 245 channels covering over eighty-seven cities were given licenses for FM broadcasting. The licence was valid for ten years.

Table 8.6   Categorization of Cities for Frequency Modulation Broadcasting Service

Category of Cities	Population Criteria for Classification of Cities
A+	Metros
A	Population > 20 lakhs
B	Population > 10 lakhs and up to 20 lakhs
C	Population > 3 lakhs and up to 10 lakhs
D	Population > 1 lakhs and up to 3 lakhs

Table 8.7   Details of the Frequency Modulation Auction

	First Phase	Second Phase	Third Phase Recommendations
Method of allocation	Single-stage auction	Two-stage auction; first stage for eligibility and second stage for financial bidding; permission shall be granted on the basis of One-time Entry Fees (OTEF) quoted by the bidders (closed tender system).	Similar to Phase II auction
No. of channels and cities	108 frequencies in 40 cities auctioned; finally only 21 channels in 12 cities became operational	337 channels encompassing 91 cities; finally 245 channels spanning over 87 cities were licensed; all cities with population > 3 lakhs	Category A+: 9 to 11 channels; Category A: 6; Category B: 4; Category C: 4; Category D: 3
Reserve price and licence fees	*Reserve Price:* A+ city: Rs 125 lakhs; A: Rs 100 lakhs; B:Rs 75 lakhs; C: Rs 50 lakhs; D: Rs 20 lakhs	*Annual licence fee:* 4% of gross revenue for each year or 10% of the reserve OTEF quoted during auction	Same as in Phase II

In order to reduce the proliferation of towers, it was made mandatory for all Phase II operators to co-locate transmission facilities in all the cities except where new towers shall be constructed by the Ministry. Every applicant and its related entities was allowed to bid for only one channel per city provided that the total number of channels allocated to an applicant and its related entities would not exceed the overall limit of 15 per cent of the total channels allocated in India. This clause was put in to reduce monopoly of the content and hence the market.

At the end of the two phases, ninety-two Indian cities were covered by private FM broadcasters, all with population > 3 lakhs. With the objective of covering cities having a population below 3 lakhs and where there was demand for radio service, the TRAI released a consultation paper in January 2008 for the third phase of FM broadcasting and made its recommendations. The majority of the recommendations were accepted by the government. A summary of the government-accepted recommendations is given in Table 8.7. The government is set to auction channels in the third phase sometime in 2011.

* * *

Starting with a sole government broadcasting arm, the broadcasting services industry in India has evolved into multi-operator cable TV service. With the guidelines for DTH ad HITS allowed, the industry is all set to witness a revolution in digital broadcasting services in the years to come. Apart from the TV story, FM broadcasting is set to make inroads into smaller town and in rural areas soon with the third phase of licensing.

9

Telecom Manufacturing, Research and Development, Software Development, and Outsourcing

The Telecom Value Chain

The telecom ecosystem comprises various players in which the service providers are just one part. The ecosystem consists of original Network Equipment Manufacturers (NEMs); handset makers; infrastructure providers such as those who install and maintain towers (also called as Tower companies), lay fibre, duct, and other passive infrastructure for the use by service providers; application and content developers who provide their digital services to the consumers either directly or through service providers; semi-conductor and Electronic Manufacturing Services (EMS) who provide silicon platforms and associated design services to NEMs or handset vendors; telecom software companies who provide middleware, embedded equipment software, testing and validation services for NEMs or handset vendors; and finally value-added retailers who assemble components and provide components such as SIMs for activation of mobile services of the associated service providers. Figure 9.1 shows the interdependence of the various stakeholders (TRAI, 2010).

History of Telecom Manufacturing in India

Before 1984, telecom service and telecom manufacturing both were fully under the DoT, Government of India. The Central public sector

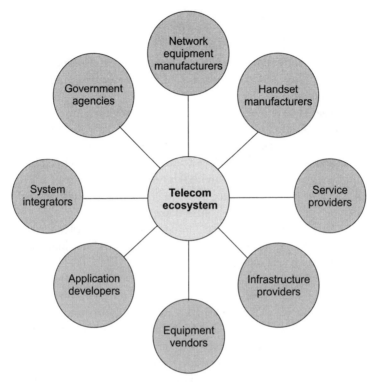

Figure 9.1   The Telecommunication Ecosystem

units (PSUs), namely Indian Telephone Industries (ITI) and Hindustan Teleprinters Ltd. (HTL) had been meeting the requirement of telecom equipment in the country. Over and above, there were Telecom Factories directly under the control of the DoT primarily to take care of the needs of telegraph line equipment. Till 1984, the switching equipment was basically the electro-mechanical types whereas the transmission equipment was open-wire carrier systems and analogue microwave systems. For crossbar type of equipment ITI had technology collaboration with Belgium which was further modified by the Telecom Research Centre (TRC) to suit the typical needs of Indian telecom. The open-wire carrier equipment and line equipment were developed indigenously and transmission equipment was developed by companies such as Bharat Electronics Ltd. (BEL) and ITI in tandem with TRC.

In 1984, the government allowed manufacturing of terminal equipment, mainly telephone instruments to be manufactured by private companies. The government shortlisted three technologies for the manufacture of Dual-tone Multi-frequency (DTMF) pushbutton telephone instruments. Only two companies, namely Ericsson and Siemens transferred the technology to a number of private companies and PSUs (mainly state-owned like Uptron, Keltron, Meltron, Punwire, and so on). This was the beginning of private-sector manufacture of telecom equipment.

As regards switching equipment, the Centre for Development of Telematics (C-DoT) was established in 1984 with the purpose of developing low-cost digital switching technology more suited for rural application. The C-DoT developed and transferred the technology to a number of private companies and PSUs for the manufacture of 128–256 and 512 Port exchanges. Later the switching capacity was upgraded to 1,400 lines and then ultimately 40,000 lines.

In 1984–5, the ITI also entered into a technology transfer agreement with Alcatel for the manufacture of the E-10B Digital Switching System. The ITI had set up a separate unit at Mankapur, UP for the manufacture of E-10B Exchanges. This agreement with Alcatel was comprehensive and Alcatel provided technical know-how for the manufacture of almost all the components of the Exchange, including software source code.

Due to the increased demand for colour TV, pushbutton telephone instruments, RAX/MAX of C-DoT and E-10B Exchanges produced by ITI, sizeable investments were made for the manufacture of electronic components during that period. Even integrated circuits for E-10B exchanges were manufactured by companies like BEL, Continental Devices, Semiconductor Electronics, and so on. As a result, a high level indigenization of up to 80–90 per cent was achieved within a short span of time.

After the formation of the Telecom Commission in 1989 and with the start of economic liberalization in 1991, the DoT took a decision to induct a few more technologies, particularly for large digital switches with internationally available state-of-the-art features with a proviso that only indigenously manufactured equipment was to be procured by the DoT.

The period starting 1991–2 exhibited an increase in the volume of traffic and thus the need was felt to induct high-capacity telephone exchanges and accordingly the DoT short-listed six technologies developed by Siemens, Ericsson, AT&T, Fujitsu, NEC, and Alcatel. As procurement was to be made from indigenous companies, except NEC all these companies set up their manufacturing bases in India either directly or through technology tie-up with Indian companies. The ITI had a tie-up with Alcatel whereas HTL had a tie-up with Siemens. But both Alcatel and Siemens also set up their manufacturing base separately in the private sector. Unlike the E-10B Exchange, the transfer of technology did not include the software portion including transfer of source code. But because of intensive training and experience, the ITI and HTL as manufacturers, and the DoT as a service provider with an in-house training programme, managed to acquire adequate skills in the operating and maintenance of these systems. Still, the software level maintenance support, of course, second-line remained with the technology providers.

Initially, by virtue of the procurement procedure in place at that time, all these companies used to get a portion of the tendered quantity, depending upon their position in the tender. However, this did lead to cartelization by the suppliers at times. The DoT later decided to put in place criteria to eliminate high-price bidders. For example, apart from the three lowest bidders, others would not be awarded any part of the order. Due to this clause, Fujitsu failed to secure orders from the DoT/MTNL due to higher bids and this virtually closed down their manufacturing operations in India.

At the beginning of 1994, the process of deregulation in the telecom sector started by inviting bids to operate cellular and basic telephone services from private operators. These licenses did not carry any mandatory requirement of buying equipment from indigenous sources (however, there was a clause for giving weightage in the evaluation of tenders for basic telephone services for using indigenous equipment). Private sector operators procured telecom equipment from foreign sources, perhaps due to the following reasons (DoT, 2004):

1. Quantitative restrictions on the telecom equipment had been progressively reduced as per India's commitment under the WTO.
2. Long-term credit by the overseas suppliers at low rates of interest. In some cases, there are also options such as converting the payments into vendor firm's equity holding.
3. Customs duty on import had been reduced progressively.
4. In 1997 Indian branches of foreign companies were allowed to carry out cash and carry wholesale trading. This facilitated foreign companies to import and supply the equipment.

During the period 1995 to 2000, both on account of the growth in the private sector telecom operations as well as deployment of newer technologies, MTNL and BSNL also followed the trend and in their tenders the clause 'Indian manufacturers' was amended to 'Indian manufacturers/suppliers'. This led to a situation wherein the procurement was made from Indian suppliers who were just importing and supplying the equipment to these PSUs. On the other hand, the private telecom operators have been importing directly from foreign suppliers.

With the change in the telecom product requirements from fixed-line switch to GSM/CDMA, the Indian manufacturers (mainly C-DoT) could not keep pace with the changing technology resulting in stoppage/closure of manufacturing units whereas multinational companies (MNCs) who had the technologies, preferred to do trading rather than technology transfer to India.

Additionally, with a view to meeting the emerging demand in the 1990s large capacities were created for Polyethylene Insulated Jelly-filled (PIJF) cables and optic fibre cables both in the public as well as private sector. One of the companies also set up an integrated manufacturing plant for manufacture of optic fibre from raw material and managed to export fibre even to Europe and the US.

The Indian Institute of Technology (IIT), Chennai developed the cor-DECT technology for providing access through WLL. After successful evaluation, the technology has been inducted in the BSNL/MTNL networks. For manufacture of equipment, the IIT through Midas Communication transferred the technology to two to three indigenous manufacturers. Some individual efforts have also resulted in the development of indigenous technologies. Tejas in Bengaluru has developed a number of high capacity optic fibre-based transmission technology solutions including carrier ethernet and dense wave division multiplexing components for backhaul networks. However, the fact is that none of the manufacturing units could migrate to the GSM and CDMA equipment being extensively used now in the Indian telecom network.

Policy Provisions

The regulatory and policy framework plays an important role in creating a congenial environment for growth of domestic manufacturing.

The NTP 1999 envisaged that 'with a view to promoting indigenous telecom equipment manufacture for both domestic use and export, the Government would provide the necessary support and encouragement to the sector, including suitable incentives to the service providers utilizing indigenous equipment'.

Further to the above, the following were the important policy provisions for telecom manufacturing:

1. No industrial licence is required for the manufacture of telecom equipment.
2. One hundred per cent FDI is permitted in the telecom equipment manufacturing sector through automatic route.
3. Telecom equipment, parts and components are freely importable (except for certain wireless equipment, where a licence from the WPC is required for import).
4. Customs' duty on telecom equipment has been decreased considerably under the Information Technology Agreement (ITA) in the WTO.

The Government has been easing regulatory policies to encourage the growth of the telecom manufacturing industry. Some major policy initiatives were taken by the government in the Union Budget for 2005–6, and these included the following:

1. Zero customs' duty on components and raw material leading to a fall of up to 13 per cent in the prices of telecom equipment.
2. Further reduction in the peak custom duty from 15 per cent to 10 per cent may enhance the efficiency and competitiveness of the sector.
3. Estimated reduction of 3 per cent in the corporate tax rate and removal of mobile phones from the 'One in Six Criteria' for the purpose of income tax to further push up the growth of mobile telephony.

The government formulated a series of programmes to support localized manufacturing including five-year tax holidays, duty-free imports of capital equipment, customs and excise benefits for both Original Equipment Manufacturers (OEMs) and Electronic Manufacturing Services (EMS) companies and other suppliers who provide components for telecom products, handsets or networking equipment. It also introduced a 4 per cent Special Additional Duty (SAD) on all imported handsets, thereby giving locally manufactured handsets a 4 per cent advantage because all components used in their manufacture could be imported duty-free.

However, even with these positive policy guidelines, the production of telecom equipment is meagre compared to countries such as China. The production of telecom equipment in India is in the range of Rs 52,000 crores per annum, the Chinese telecom equipment maker Huawei Technologies alone reports a revenue almost double (TRAI, 2010).

The following factors are impacting the growth of telecom equipment manufacturing in India:

1. Private sector service providers have no compulsion to use equipment manufactured by indigenous companies. Their procurement of equipment is dependent on the choice of technology and funding mechanism, with long-term low interest credits by foreign suppliers.
2. The C-DOT and other Research and Development institutions could not develop new technologies, resulting in the closure of units set up for the manufacture of their earlier products due to a decline in demand.
3. The government allows trading of telecom equipment to foreign companies under 'cash and carry wholesale trading'. Institutional sale is considered under wholesale.
4. With the rapid growth of wireless access, GSM and CDMA, the entire demand is being met through import. The C-DOT failed to build robust GSM/CDMA products for local consumption.

The following cases in Box 9.1 illustrate the wavering fortunes of the Indian Telephone Industries (ITI) and the changing shape of India's telecom manufacturing capabilities.

Manufacturing in India Post 2004

Foreign multinationals never showed interest in setting up manufacturing units in India until 2005, when Elcoteq, the world's leading EMS vendor in the communication space set up a handset manufacturing unit in Bengaluru. The cited reasons for the newly found interest in setting

---

Box 9.1   A Decade in the Case of Indian Telephone Industries

The ITI is the oldest public sector enterprise and the largest telecom manufacturing monolith in the country, conceptualized and started in the year 1948 as the first public sector company after independence. The ITI has six manufacturing units, including three in Uttar Pradesh (Mankapur, Rae Bareli, and Naini), one each in Srinagar in Jammu and Kashmir, Palakkad in Kerala and the main facility in Bengaluru. Started as a supplier of landline telecom infrastructure equipment for MTNL and DoT, seeing opportunities in broadband and mobility, and the falling demand in fixed line, the ITI diversified its portfolio of offerings to include SDH fibre optic terminals, GSM, WLL and TDMA infrastructure devices. It also got into products like info-kiosks, IP-based PBX, and network management systems. The years 2002–4 were bad years for the ITI due to the shrinking market demand for domestic telecom products. The ITI's revenue from manufacturing and equipment was just 30 per cent of the revenue in 2003–4. The majority of the revenue was from services and a minority from turnkey solution provisioning.

To strengthen its equipment portfolio, the ITI entered into number of partnerships with a number of multinational equipment vendors as shown in Table 9.1. In most of the cases, the ITI became a trading partner for these multinationals with not much indigenization of technologies. The reason cited was the C-DOT's failure in GSM and CDMA and hence the need for technology transfer from multinationals.

The financial year 2004–5 saw the UPA government pumping in Rs 1,001 crores in a last-ditch bid to revive the ITI, the old telecom equipment warhorse. The Cabinet Committee on Economic Affairs approved a plan to revive the ITI by infusing Rs 200 crores' equity, Rs 458 crores for voluntary retirement scheme, Rs 50 crores for capital expenditure, and Rs 93 crores to clear provident fund dues in the current financial year. It was also decided to write off interest and penal interest of Rs 23.67 crores.

The ITI managed to increase its sales to Rs 1,505 crores in the financial year 2004–5 from Rs 1,278 crores in the financial year 2003–4. However, it still incurred losses as price cuts, excess manpower, and huge overheads affected its margins. Also, it failed to keep pace with technological advancements and lost market to newer players. One of the major achievements of the ITI during 2004–5 was an order worth Euro 65 million from BSNL. The order was for supply and installation of GSM lines for BSNL using Alcatel technology. Apart from domestic presence, the ITI diversified into other markets, the first of which was in Afghanistan in 2004–5 to deploy network equipment for Telecommunications Consultants India Ltd.

In 2005–6, the firm continued its upward swing with about 30 per cent increase in revenue. The Bengaluru plant of the ITI topped the high-grosser list with the highest turnover, followed by the plants in Mankapur, Palakkad, Naini, Rae Bareli, and Srinagar. Turnkey projects became an important revenue earner for the ITI. The ITI also made inroads into VSAT equipment installation and commissioning in the Ku-band with a strategic alliance with BSNL. However, the revenue declined substantially in 2007 and 2008 with the result that on 25 March 2008, the ITI declared itself sick under the Sick Industrial Companies [Special Provisions] Act, 1985.

However, in the year 2009–10, the manufacturer took a giant leap, recording a 171.8 per cent growth with a record turnover of Rs 4,732 crores against Rs 1,741 crores in the previous year (*Voice and Data*, 2010). This is the first time since liberalization that the ITI's revenue grew at this rate. During 2009–10, the ITI established a manufacturing facility for the gigabit passive optical network (GPON) at its Rae Bareli plant in Uttar Pradesh and successfully supplied equipment to the tune of Rs 270 crores to BSNL. The ITI became the first company in the country to get a Technical Approval Certificate (TAC) for the GPON products. The company also implemented successfully the 'National Population Register' project of the Ministry of Home Affairs at its Palakkad plant in Kerala. The ITI–Huawei combination enabled BSNL to launch 3G services in the south zone. In an effort to save the sinking company, ITI initiated actions for the establishment of a joint venture company at its Rae Bareli, Naini and Bengaluru plants for the manufacture of WiMax, GPON and NGN IP systems.

The company set up a Rs 77-crore data centre, which will provide customer-centric services to the government and private enterprises at Bengaluru, in collaboration with the Mumbai-based leading IT infrastructure and services firm Trimax Datacenter Services. The fifty-seat ITI data centre will offer a range of services, including co-location services; hosting services; and managed services such as server management, network management and security management. The prime business focus of the facility will be value-added services such as messaging solutions, videoconferencing and remote infrastructure management.

As per the agreement, 82 per cent of the revenues generated by the data centre will be credited to Trimax, while the rest will accrue to the ITI. The collaboration is targeting Rs 50 crores in the first year (financial year 2009–10) of operations; and is in the final stages of discussion with RailTel Corporation of India, Karnataka Police Housing, Business Intelligent Technologies, Glodyne and Madhucon to engage them as customers for data centre services. As part of the ITI's diversifying strategy, it plans to set up two more data centres at Lucknow and Palakkad.

The figure below gives the track of revenues of the ITI. Table 9.1 shows the various partnerships of the ITI.

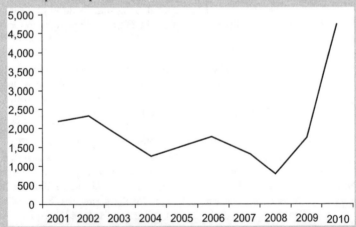

Figure 1   Indian Telephone Industries' Revenue (in Rs crore) Movement over the Decade

Table 9.1   Indian Telephone Industries' Partnerships

GSM and ADSL	Alcatel, France
CDMA	ZTE, China
corDECT	Midas, India
SDH	Tejas Networks, India
Data centre	Trimax, India

up manufacturing plants in India, especially for handsets included the exponential growth of mobile services in India, the need to be near the market and engage in localization of products, and the availability of skilled manpower at economical wages.

Elcoteq produces both terminal products and network equipment. The terminal products include high-volume wireless communication products, including mobile phones and accessories, smart phones, Personal Digital Assistants (PDAs) and accessories, wireless modules, and home communication products.

Following Elcoteq, Flextronics (now called as Aricent), the leader in EMS and wireless equipment manufacturing set up a facility in the Sriperumbudur Industrial Park in November 2006. The State Industries Promotion Corporation of Tamil Nadu (SIPCOT) Limited, a fully government-owned premier institution, has been a catalyst in the development of small, medium, and large-scale industries in Tamil Nadu. The Tamil Nadu government set up an industrial park in Sriperumbudur, about 30 km from Chennai to reap the advantages of manufacturing in India, which included access to skilled talent and lower wages and operational costs. The Flextronics unit is modelled after Flextronics' fully integrated, high-volume facilities in other countries, including Brazil, China, Hungary, Malaysia, Mexico, and Poland. These industrial parks provide total supply chain management by co-locating its manufacturing and logistics operations with its suppliers at a single low-cost location.

In January 2006, Nokia announced plans to set up a manufacturing base in India with a commitment to invest US$ 150 million over a four-year period. This was Nokia's fifteenth manufacturing facility, and one that would manufacture both handsets and network equipment. By December 2010, this facility crossed production volumes of 350 million handsets. The facility also added another milestone to its local operations by starting exports to North America and Europe, taking its total export markets from India to seventy. The facility also had the distinction of achieving the fastest ramp-up across all Nokia's manufacturing plants globally. The current production is supplied to both the domestic market as well as exported to countries in the Middle East and Africa, Asia, Australia, and New Zealand, besides North America and Europe.

The other major players that have set up manufacturing facilities in India include: the Swedish major Ericsson for GSM radio base stations' manufacturing facility in Jaipur in 2005; and South Korean conglomerate LG Electronics for GSM mobile phone plant in Ranjangaon, about 55 km from Pune in September 2008. The government has set up the Telecom Equipment and Service Export Promotion Council, and the Telecom Testing and Security Certification Centre to further the manufacturing of telecom products in India.

The status of the telecom equipment production and exports is presented in Figure 9.2 (TRAI, 2010). As compared to this, the Chinese telecom manufacturing giant Huawei alone reported revenue of Rs 82,350 crores in 2008. Hence the telecom manufacturing industry in India has still not reached a scale of operation in correspondence with growth of the telecom services experienced in the country.

Table 9.2 compares India and China on the factors that impact telecom manufacturing (*Voice and Data*, November 2009):

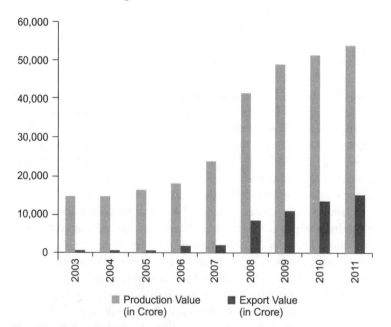

Figure 9.2   Status of Telecom Equipment Manufacturing in India

Table 9.2 Comparison of the Telecom Manufacturing Environment in India and China

Factor	India	China
Manufacturing history	Started only in 2004	Long
Growth of domestic market	High	Medium
Foreign investment	Poor	Very high
Supporting infrastructure such as electricity	Poor	Very good
Government support	Nil; zero customs duty on imported equipment discourages indigenous manufacturing	Very good; included preferential credit facilities for firms
Cost of capital	High	Low due to support from government and financial institutions
Trained manpower in manufacturing	Less	High

However, of late, the Indian companies have started making significant progress in the handset space. In the financial year 2009–10, the market share of Nokia in India stood at 52.2 per cent, a good 11.8 per cent fall from 64 per cent in the previous financial year. On the other hand, the market share of the Indian handset makers which, combined, was between 3–4 per cent in the previous financial year, has grown up to 14 per cent. Micromax mobiles topped the chart (of Indian makers) with 4.1 per cent share followed by Spice (3.9 per cent) and Karbonn mobiles (3 per cent). Recent entrants such as Lava mobile, Lemon mobiles and Max mobile too are hovering around the 1 per cent market share mark. The growth of Micromax and its astounding success against multinationals such as Nokia is detailed in Box 9.2.

Research and Development in Telecom

Research and Development (R & D) is an important link in the telecom supply chain which has the potential of bringing immense value to the country (TRAI, 2010). For a manufacturing segment, especially in telecom where the time to market is shrinking fast, R & D and innovation are important prerequisites. In general in India, stress has been less on R & D and innovation, creation of Intellectual Property Rights (IPR), and manufacturing high-technology products. Until now, the country

Box 9.2   The Micromax Story

Micromax started with the trading of mobile handsets imported from China and Taiwan and transformed the domestic handset market. With a well-defined product vision and a Research and Development (R & D) set-up to support it, Micromax has been successfully generating innovative handsets that have revolutionized the telecom consumer space. Through its thorough understanding of the Indian market, Micromax is coming up with innovative handsets with unique features to compete in the market. Some of the features Mircomax is implementing through its R & D efforts in its handsets are:

- Phones that can also be used as a remote control for consumer durables in a household, such as TV, air-conditioning units, and a DVD player (called a Universal Remote Control Mobile Phone).
- Single-click access to popular social networking sites like Facebook and Twitter.
- Clamshell phones containing Swarovski crystals on the keys and priced at just Rs 5,500 targeting women in smaller towns, who would like the chic appeal of this handset.
- The 30-day standby battery life on mobile sets targeted at rural areas where power failures are common (called Long Life Battery Mobile Phone).
- Unique dual-SIM capability in about 85 per cent of its handsets to enable prepaid subscribers to use more than one connection for plan price, roaming location and other such associated factors.
- Handsets with capabilities to switch between GSM and CDMA networks options
- Mobile handsets with three-dimensional motion gaming technology that target the youth (also called Gaming Mobile Phone).

Thus it is not only the lower price points but also the growing R & D capabilities that have helped Micromax achieve an increasing market share.

has demonstrated significant innovation capability in only two kinds of hardware (Mani, 2005). The first one is in small and large electronic digital switching systems, known popularly as the C-DOT digital switches and the second one is in a WLL access technology known as CorDECT (see Boxes 9.3 and 9.4).

While these isolated cases illustrate the potential for R & D in telecom in India, the following challenges have presented in TRAI (2010) for the poor under-explored space:

1. Analysis of the value chain in telecom indicates that stress had been more on the assembly of low-value products resulting in lower

Box 9.3   The C-DOT Story

It was in August 1984 that the Centre for Development of Telematics or C-DOT was set up with the specific intention of indigenizing digital switching technology to meet India's unique requirements. The C-DOT was established as an independent society to help develop a series of digital switching products to meet Indian requirements. It was started by Dr Sam Pitroda under the aegis of the then Prime Minister Mr Rajiv Gandhi. The concept was to build indigenous Indian technology for digital switching systems with focus on rural connectivity, accessibility as opposed to telephone density, ancillary industries and young talent with new energy and new work culture, work ethics, work norms and work values. The overall strategy was to design, develop, and manufacture products to suit Indian climatic conditions, especially for rural exchanges without air-conditioning at substantially lower costs and at the same time train human resource in ICT to manage, maintain and develop products for the future.

The C-DOT made a public commitment to develop products in thirty-six months for an investment of Rs 36 crores. As a result of openness, transparency and public accountability the C-DOT was fortunate to get substantial support from the national and local media. Right from day one, the C-DOT focused on public–private partnership with a clear understanding that the products designed by the C-DOT will be manufactured by public and private companies. The first product was a small rural exchange to connect villages. The small-capacity Rural Automatic Switches (RAX) with up to 256 terminations or ports were designed and developed in-house by the C-DOT. It is an easy-to-install and fault-tolerant system with inbuilt redundancy. Besides requiring no air-conditioning, it withstands wide temperature fluctuations ($-20°C$ to $50°C$) and humidity. Moreover, it consumes very little power. A distinguishing feature of the C-DOT RAX is its simple and flexible connectivity through a wide range of transmission systems such as Ultra High Frequency (UHF), Very High Frequency (VHF), radio and satellite. As it is program-controlled, it can be easily adapted to different network requirements through software changes. Thereafter a small PBX was delivered for the business community.

Then came Digital Switching System (DSS) Main Automatic Exchanges (MAX), a family of digital switching systems which offers a total switching solution for national telecom networks. The C-DOT first developed a medium-sized 2,000-line digital exchange; then a 16,000-line exchange; and eventually a 40,000-line large exchange to meet urban needs. The C-DOT DSS MAX products have the proven ability to serve as local, toll, transit and Integrated Local-cum-transit (ILT) switches. The modular architecture of the exchanges ensures cost-effectiveness and protection of investment as the demand in a service area grows. It is possible to begin with a small switch in a potentially high-growth area and to increase exchange capacity as the demand grows, simply by adding more modules to the existing exchange. Over 20 million lines of the C-DOT exchanges are in service in the country. The rural exchanges were used to provide STD/PCO to improve access to telephones nationwide.

The C-DOT redefined its roadmap on its twenty-fifth anniversary (*Voice and Data*, 2009). The organization has realigned its efforts, with a focus on the four major directions, keeping in view the relevance and need in the present scenario. The major developmental schemes in these directions are projects of national and strategic importance for rural areas through shared GSM RAN. The organization is focusing on projects for software-intensive telecom solutions in the form of network management systems (NMS), which provide a common umbrella of management of a network equipped with products from diverse vendors. The C-DOT is also undertaking futuristic study projects like 'One Number', which will aim to utilize a unique number like a social security number for a mobile personal number, cognitive radio and advanced optical network technology. The organization plans to undertake a new project on a central monitoring system, which will be deployed across the networks. The organization has been chosen as a nodal agency for the implementation of the central monitoring system.

In a bid to push telecom R & D and manufacturing in India, the C-DOT and French telecom major Alcatel signed an agreement in 2005 to set up a global R & D centre for new broadband access technologies in Chennai. Further, to lead R & D and manufacturing in wireless broadband, the C-DOT and Alcatel set up a joint WiMax Reality Centre in 2007 (*Voice and Data*, 2007). The WiMax Reality Centre is said to be the first of its kind in India to focus entirely on IEEE 802.16e (Rev-e) standard, which replaces 802.16d (Rev-d)) and is far superior. The new standard promises more performance and is agile enough for fixed WiMax deployments. In consonance with that the Centre was set up to design and develop ready-to-industrialize products and reference designs. The Centre is equipped with complete architecture to leverage 802.16e and will unravel its potential in areas like VoIP, high-speed broadband Internet and applications like IPTV, video streaming and mobile TV. In 2010, the Centre set aside an outlay of Rs 269 crores for the R & D needs of the C-DOT.

However, the C-DOT faces stiff competition in telecom R & D and manufacturing due to the recent entry of Chinese manufacturers such as Huawei and ZTE. With Indian private operators tilting more towards LTE, the investments of the C-DOT in WiMax have to be justified.

A detailed analysis of its development and subsequent manufacture can be found in Mani (2005). However, as Mani (2005) indicates, the C-DOT did not strategically view mobile technology as the future of communication. Its lack of market foresight on developing GSM and CDMA-based equipment resulted in foreign multinationals filling up the local demand through imports and minor customization. Of late the new leadership has been pushing forward R & D in the wireless space through a number of initiatives some of which are listed above, the results of which are yet to be seen.

accruals to the Indian companies. Stress has been less on R & D and innovation, creation of IPR, creating a component supply chain, and developing and manufacturing high-tech products.

2. Telecom product development requires heavy investment, sometimes with indefinite outcomes. It is felt that there are no meaningful R&D grants that are available from the Government. Other countries provide liberal R&D funds as grants or low-interest loans for developing telecom products.

3. Presently, the demand for telecom equipment is largely met through imports which could be the reason for meagre R&D work within the country.

There are some exceptions to the above as indicated by the work of Tejas Networks, started by the Sycamore-fame Gururaj Deshpande in 2000, in developing software-based intelligent optical networking products that provide high price/performance in their class, enabling carriers to maximize revenue-generating services while optimizing their overall network costs. Currently, most of the BSNL and MTNL networks have deployed the optical networking solutions of Tejas in their backbone networks (Mani, 2005).

Telecom Research and Development Outsourcing Services in India

Of the total US$ 8–9 billion telecom engineering R&D outsourcing market, India has about 30 per cent of the market share at US$ 2.4–2.6

Box 9.4 Contributions of IIT-M in Telecom Manufacturing and Research and Development

The Telecommunications and Networking (TeNeT) group at the Indian Institute of Technology Madras (IIT-M), under the leadership of Professor Ashok Jhunjhunwala, has been playing a key role in defining and developing access technologies suitable for India. As the cost of backbone network and switch costs reduced considerably, the emphasis, especially in countries such as India with a substantial population having lower disposable income levels is to shift to low-cost wireless access technologies. The cost advantages of wireless access systems over wire-line networks are explained in Chapter 2. The IIT-M along with Midas Communication Technologies, Chennai, and in partnership with Analog Devices, USA took up the development of a Digitally Enhanced Cordless Telephony (DECT)-based WLL system in the late 1990s.

The system, referred to as corDECT, has an interesting architecture and consists of (i) a fixed part consisting of a DECT interface unit acting as a 1,000-line wireless switching unit providing V5.2 interface and connected to the main exchange and (ii) a weather-proof Compact Base Station (CBS) that is connected to the DECT Interface Unit (DIU) on three pairs of copper wires carrying signal and power and (iii) the wall mountable subscriber terminal (called Wall Set–WS) with a roof-top antenna providing a line-of-sight to the CBS. The WS has an interface for a standard telephone and an additional RS232 interface for a computer, enabling digital Internet connection between the WS and DIU. The Internet packets are routed through circuit-switched digital end-to-end network connection all the way up to the local Remote Access Switch (RAS) after which it is bifurcated and sent to the ISP network using a leased line connection. This reduces the demand on the main exchange and provides affordable relatively higher speed Internet connectivity to end-subscribers. All subsystems are built primarily using digital signal processors with soft solutions, cutting down moving parts considerably, providing the needed flexibility in configuration and reducing the per line cost. The technical details of corDECT are presented in Jhunjhunwala *et al.* (1998). Figure 9.3 gives an illustration of the corDECT system deployed at the customer premise.

The system was adopted especially for rural connectivity by Shyam Telecom, a private operator in Rajasthan; and the state-owned BSNL in Maharashtra, Madhya Pradesh, and Andhra Pradesh. It was also selected by UNDP as the cheapest and fastest mode of Internet connectivity in rural areas of the country. However, higher penetration of mobiles, associated with reduced cost of deployment as specified in Chapter 3, has stalled the development of corDECT in many parts of the country.

WS/WS – IP : Wallset with external antenna

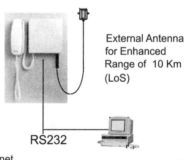

External Antenna
for Enhanced
Range of 10 Km
(LoS)

RS232

35 kbps/70 kbps Internet
traffic plus simultaneous voice

IIT Madras

Figure 9.3   Photograph of the Wall Set of the corDECT System

billion. The NASSCOM and Booze report estimates that the total off-shoring of telecom R & D services is expected to reach US$ 5.5–6 billion in 2020 and will continue to be one of the anchor verticals (NASSCOM, 2010). The domestic market is also expected to grow from the current US$ 50–100 million to US$ 800–900 million in 2020. Broadly, the telecom engineering R & D services can be categorized as follows:

1. Network equipment manufacturing: Engineering R&D services for RAN, core networks, emerging 3G/4G network components, management support system and operational support system development, Femto cell and fixed–mobile convergent networks belong to this category. Major outsourcing multinationals include Nortel (now called Kapsch), Alcatel Lucent, Ericsson, Nokia Siemens, Motorola, Huaweii, and ZTE. Almost all of them have their captive R & D centres in India. Third-party vendors include large companies such as Wipro, Infosys, TCS, and Cognizant Technologies who have separate telecom verticals and niche firms in this domain such as Sasken, Aricent, and Mindtree.
2. Mobile handsets: Protocol (GSM/LTE/WiMAX), middleware, applications development, and connectivity solutions (Universal Serial Bus [USB]), Bluetooth, near-field communication) for multinationals such as Nokia, LG, Samsung, Sony Ericsson are being provided by Indian software companies. The biggest change witnessed in this area is that proprietary mobile operating systems such as Symbian, S40 and S60 of Nokia are being replaced by open-source Android developed by Google. This has disrupted the market for outsourced services. Taking advantage of this move, some of the niche players in this field such as Teleca and Persistent Systems have penetrated the market.
3. Integrated Chip (IC) Design: Design, synthesis, validation and testing of IC and systems on chip (SOC) solutions for mobile handsets for multinationals such as Texas Instruments, Qualcomm, Infineon, and Broadcom are being provided by the Indian R&D outsourcing companies.

Though cost arbitrage is one of the primary reasons for the growth of telecom R & D outsourcing to relatively lower cost countries such as India, following are the other reasons:

1. Access to a huge number of engineers as resources for the projects.

2. Growing demand for locally relevant innovation in India and hence the need to be nearer to the market.

The article in Box 9.5 (Mody and Sridhar, January 2011) gives a view of what to expect in the telecom space in the years to come.

One of the niche players in the telecom R & D services space is Sasken Communication Technologies that carved a mark for itself through high-end product line management and intellectual property creation. Box 9.6 gives the story of Sasken.

Box 9.5   Welcome to the Changing World Order in Information Technology and Telecom

Much like the financial crisis that precipitated a new world order, a quiet revolution of some sorts is happening in the telecom industry worldwide. The bankruptcy of stalwarts such as Nortel and the impregnation of Google and Apple into the mobile phone space at an amazing alacrity are changing the world order once dominated by the likes of biggies such as AT&T. What are these changes and what can we expect in the future?

The world renowned Bell Labs, which boasted of seven Nobel prize-winning inventions, was reduced to almost nothing when the Alcatel-Lucent combine pulled out basic science, material physics, and semiconductor research due to market and financial pressures. The last Nobel prize-winning work which came out of the Bell Labs was in 2009 for the work of Willard Boyle and George Smith for their invention on the imaging semiconductor circuit, the Charge-coupled Device, in the late 1960s. The last Nobel laureate at IBM was Harry Markowitz in 1990 for his co-invention of economic theories on stock market risks and reward. Is there a tectonic shift in the areas of corporate research and development?

Though breakthrough inventions such as the above have declined of late, the time-to-market of products has decreased considerably, leading to smart innovations in marketable areas including networking, high-speed electronics, wireless networks, nanotechnology and software. While IBM has mastered the art of being ahead of others especially in IT services, most of the large organizations such as Nortel and AT&T faltered while adapting to the changing market conditions. The new world order is to be nimble and agile in producing solutions that can survive in the extremely competitive marketplace. Though 'multi-touch' was invented way back in 1991 by Pierre Wellner as explained in his classic paper on the 'Digital Desk', Apple showed the way to package multi-touch on mobile phones and be commercially successful with the launch of iPhone in 2007. Similarly, with customized and flexible offerings, MediaTek of China has risen speedily to the second place in the mobile handset chipset business once dominated by the likes of Texas Instruments.

The other phenomenon that is taking the mobile world by storm is the community-based application development on open platforms. Google transformed the way in which applications are built and distributed. Though the open-source software movement just did not take off in the PC world, thanks to the close control of the market—both of consumers and developers—by Microsoft, things are very different in the mobile space. Google's Android is being willingly accepted by all stakeholders including developers, handset makers and consumers as the default operating system for mobiles as well as consumer electronics. The proprietary platforms are withering. The huge direct and indirect network externality effects of open platforms and operating systems have helped Google make giant strides in this space in which it did not have any presence at all a couple of years back. The order here is 'collaborative innovation'. Firms do not have to invent on their own from the basics. The 'flat world' has opened up avenues for tapping the innovative potential of millions of individuals and small firms across the world and garners them for competitive advantage. This opens up avenues of growth for R & D outsourcing firms in India. However, because it is open, innovations are very rapid and have a very short lifecycle. Firms in this space need to be extremely smart in forecasting business and technology landscapes, making appropriate investments and building capacities. Extrapolating the past is similar to driving a car by looking at the rear-view mirror. You never know what will hit you in the road ahead. Hence one has to be conscious of the ever-evolving market environment to make sure that one bets on the right products and services.

The third intriguing aspect in research and development is the role of software patents in this emerging scenario. In 2007, IBM received 3,125 US patents from the US Patents and Trademark Office. This was the fifteenth consecutive year that IBM received more US patents than any other company in the world. However, an overwhelming majority of the 700 software entrepreneurs who participated in the famous *2008 Berkeley Patent Survey Report*, ranked patenting as the dead last amongst the seven strategies for attaining competitive advantage. Pamela Samuelson in a recent issue of the *Communications of the ACM*, points out that the new-age software firms regard 'first-mover advantage' as the single most important strategy for attaining competitive advantage. The more open platforms and the increased cost of obtaining and enforcing patents are barriers to software patenting. This requires that the firms be agile, niche, and be business-savvy to survive in the competitive world of research and development, especially in software.

The fourth change is to participate in the local market. Traditionally, Indian firms have been supporting R & D needs, especially in telecom, of multinationals that were developing products for mature markets. Not being nearer to the market was always a disadvantage as the requirements could not be completely conceptualized and fulfilled. However, the recent strides of India in telecom and the imminent deployment of 3G and BWA services by

Indian telecom companies provide opportunities for Indian R & D firms to provide customized offerings to local markets. Examples abound as to how India and other emerging markets have become a very important market for multinationals such as IBM, Nokia and Vodafone.

The above changes open up opportunities for Indian R & D software services firms. However, the challenge is to develop the required competency, foresee technology and market evolution, and react with speed.

Innovation in Network Management Services

While telecom manufacturing did not grow, telecom services have shown robust growth, often leading to the development of some of the best practices in the industry. The rapid growth of mobile services brought innovation in service offerings.

One of the main reasons for telecom companies to offer such low airtime charges and still make profit is due to extreme cost management practices. The strategy adopted by most of the telecom companies to manage costs better is to outsource network management services.

The following three different areas of outsourcing are being pursued by the telecom operators:

1. Network operations and management: Network capacity planning and deployment, integrating network equipment, network maintenance, installation and fault repair, and deployment of new network services.
2. Information technology management: Information technology infrastructure management, desktop and server management, Operational Support Services (OSS), Billing Support Services (BSS), customer relationship management software development and implementation, deployment of new IT services.
3. Customer relationship management: Call centre services.

However, the nature and complexity of the above operations differ across different types of services (namely, mobile, fixed) offered by the telecom operators. The amount of outsourced work differs depending on the models adopted by the operators ranging from full outsourcing to managing capacity, details of which are given in Sridhar (2009). Box 9.7 provides an example of such an innovative outsourcing model pioneered by Bharti Airtel in 2003.

* * *

Box 9.6 Sasken: A Niche Player in the Telecom Research and Development Space

Sasken Communication Technologies (www.sasken.com) is a focused wireless communication software company with corporate headquarters in Bengaluru, providing research and development outsourcing services. The core expertise of Sasken is in developing embedded communication software for companies across the communication value chain, namely semiconductor vendors, terminal device vendors, network equipment manufacturers, and operators. Global Fortune 500 Tier 1 companies in each of these segments are part of Sasken's customer profile. With an annual revenue of about Rs 575 crores (at March 2010) and more than 2,800 employees worldwide, Sasken is recognized in the market for its contributions to the country's high-end engineering R & D services outsourcing market.

Started by Rajiv C. Mody, currently the Chairman and CEO, in a garage in Silicon Valley as an Electronic Design Automation tool company, Sasken diversified into related areas in the wireless communication space over the past decade. Sasken also had a strong and unique presence in telecom software product development when most of the IT companies in India focused on services. More than fifty phone models and 50 million phones carried Sasken's intellectual property in communication protocols, multimedia engines and applications. However, currently, in tune with the market demand, the services contribute to much of its revenue.

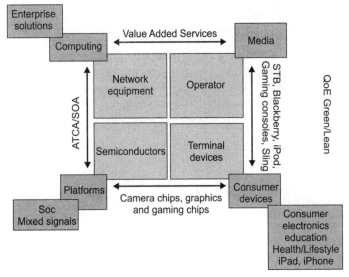

Figure 9.4   Portfolio of Business of Sasken

Sasken had its headquarters in Bengaluru, India with development centres in Pune and Chennai in India and wholly-owned direct subsidiaries in China, Finland, Japan, and the US. Sasken also created near-shore development centres in China and Finland, to provide customer-centric solutions in the respective geographical areas. Through an acquisition in Finland Sasken also diversified into hardware design and testing apart from software services. Sasken followed the butterfly model given in Figure 9.4 for its product and services portfolio.

The core of Sasken's offerings includes semiconductors, terminal devices, network equipment and operators. However, as indicated by the movement towards commoditization of wireless software, especially in handset space, Sasken intends to expand the core offerings as shown in the figure. For example, as Android open-source operating system is becoming the *de facto* standard in not only mobile handsets but also in consumer electronics and automotive infotainment, the natural expansion choice for Sasken is to use its expertise in Android to build products and services for consumer electronic devices such as set-top boxes and digital video recorders. With mobile service providers searching for content and VAS to enhance their ARPU, there is an opportunity for Sasken to be involved in VAS applications/content creation, deployment and sustenance. Sasken is also actively looking to develop cheaper network equipment, especially in the areas of 4G as a suitable alternative for the Indian and other emerging markets. The depth of Sasken's offerings in semiconductors, especially in chip design, and verification allows it to explore providing end-to-end solutions starting from chip to applications in the wireless communication services space to semiconductor companies, network equipment firms, handset vendors and mobile operators.

Using these capabilities and its intellectual property, Sasken created a satellite phone (iSatPhonePro) for the International Maritime Satellite (INMARSAT) Company in 2010. This was the first ever attempt by an Indian company to design, develop and test a full satellite phone with all of its components. The phone had more than ten vendors supplying hundred different-ent components, with about 150 engineers and four project managers working for eighteen months. Colorado division of Emergency Management in the US became the first customers to use IsatPhone Pro following its commercial launch in June 2010. The development was done in a multi-site mode with members of teams working across Sasken development centres in Bengaluru, India; Kaustinen, Finland and Bochem, Germany. Detailed and thorough planning, well thought out risk mitigation strategies, and determined execution saw Sasken cruise through successfully in this US$ 14 million project for INMARSAT.

Box 9.7   Outsourcing Success Story of Bharti Airtel

In 2003, Bharti Airtel, the largest private telecom operator in India outsourced its network management and IT operations. In March 2003, Bharti outsourced its data centre operations, billing support systems, application development, and customer relationship management valued at US$ 750 million in a ten-year contract with IBM (Sridhar, 2010). Bharti's agreement with IBM was based on revenue-sharing. Over a period of ten years, IBM would design, build up, and maintain Bharti's IT network in a full outsourcing model in exchange for a portion of Bharti's revenues (Martinez-Jerez and Narayanan, 2006). During the contract period, IBM would operate Bharti's data centre, its disaster recovery site at Chennai, India, and the billing that Bharti does in its LSAs around the country, its CRM programme, all applications' development, as well as the IT helpdesk. IBM would also handle over 80 per cent of Bharti Airtel's current programme and project management.

Bharti followed this with outsourcing network management to Ericsson (in fourteen LSAs) and Nokia Siemens (in seven LSAs) for a three-year US$ 724-million contract. During this period, Ericsson and Nokia Siemens would manage base stations (antennae, switches, routers, transmitters and receivers) within their areas of operation, deploy new base stations as required, roll out new networks and applications, and take on board roughly 800 of Bharti Airtel staff. They would also add new staff dedicated to Bharti Airtel as the network and business of Bharti expanded. As a result of these outsourcing contracts, Bharti Airtel would now just handle a few things: marketing, sales and distribution. Apart from that, it would just monitor its vendors, see that they stick to the parameters of the contracts, and make sure that they deploy and build only the best systems and networks for Bharti's operations (Singh and Dubey, 2004). While the IBM outsourcing deal followed the full outsourcing model, the network management was on a managed capacity model. The success of this outsourcing model is explained in detail in Martinez-Jarez and Narayanan (2006).

The managed capacity model with network vendors allowed Bharti the flexibility to pay the vendors on Erlang basis as and when the network capacity was up and running and was used. This avoided excess capital expenditure by Bharti on its network infrastructure deployed in anticipation of future demand. This also allowed Bharti a window of credit period and hence rationalized the capital expenditure in tune with the actual demand. The focus of the Indian mobile operators was switching from plain vanilla voice to VAS. Indian operators had to adopt 2.5G and 3G services to provide these services which required large capital expenditure (Martinez-Jarez and Narayanan, 2006). The managed capacity model followed by Bharti protected it against such large expenditures. Due to capacity-based payment to network vendors, it was estimated that the capital expenditure of Bharti declined by as much as US$ 730 million in 2007 (Singh and Dubey, 2004).

However, outsourcing entails contractual risks such as vendor bankruptcy, vendor's inability to deliver, and contract breach by the vendor. To mitigate these, the firm engages in screening the vendor ex ante and monitoring vendors ex-post. In screening, the firm's goal is to identify a potential vendor who is best suited for providing a certain service. Through proper screening, firms can reduce the risk of opportunism and the adverse selection problem is mitigated. By outsourcing to select large expert firms such as Ericsson, Nokia Siemens and IBM, Bharti reduced the vendor selection risk. By having a comprehensive Service Level Agreement (SLA) with over 100 metrics and monitoring them closely Bharti mitigated ex-post risks.

Following the above, Bharti in 2009 also outsourced its fixed-line services to Alcatel Lucent. Fixed-line services are much different from mobile services as it involves coordinating with civic authorities for digging and laying down the copper lines, entering into homes to fix and maintain jacks and modems, and provide face-to-face customer service. Following the above, major telecom operators in India have recently handed over large IT and network outsourcing contracts to the same set of vendors. Idea Cellular signed an IT outsourcing contract in 2007 with IBM for US$ 500 million spread over seven years. Vodafone did the same with IBM for US$ 1.5 billion. Aircel, which is owned by Maxis Telecom of Malaysia followed suit with outsourcing its IT services and business processes to Wipro, an Indian IT company (Sridhar, 2009). Similarly, the network management of Idea went to Ericsson while that of Vodafone was bagged by Nokia Siemens.

Manufacturing of sophisticated telecom equipment requires many electronic components. As indicated in this chapter, domestic manufacturing and R & D of telecom products both by the multinationals and indigenous manufacturers have fostered the growth of the industry to some extent; not all components can be sourced locally and are procured through global suppliers. The ecosystem comprising the production of critical components such as integrated circuits and other sophisticated subassemblies is not well developed in India and all are imported. Non-availability of indigenous components is a major constraint facing the Indian telecom manufacturing industry, unlike the fully developed Chinese industry (TRAI, 2010). The government support for telecom manufacturing and R & D efforts is expected to give a fillip to the Indian manufacturers to compete with the low-cost, function-rich equipment supplies from Chinese vendors. The TRAI in 2011 recommended that Indian manufactured products and Indian products be given preferential market access by the government and government licensees (both private and public operators) starting from 30 per cent in 2012–13 increasing in steps up to 80 per cent by 2020 (TRAI, 2010). It has to be seen whethet these can be successfully implemented and will give the required push for telecom manufacturing in the country.

10

Partnerships and Convergence in Telecom
An Evolving Story

In the previous chapters we discussed the various segments of the service provider industry. However, there are many other industry segments that work closely with the service provider industry in the telecom value chain. We expand on the framework provided by Grover and Saeed (2003) to define the telecom value chain. The resultant categories with examples are shown in Table 10.1 (*Voice and Data*, June 2008).

Except in certain segments such as telecom cabling, telecom software, network security, the Indian telecom manufacturing sector has not grown in tune with service provisioning.

Apart from the above, MVAS providers, content providers content aggregators, submarine cable operators, and wholesale bandwidth providers as explained in some of the previous chapters, also contribute to their portions of the telecom value chain.

Types of Partnerships in Telecom

Grover and Saeed (2003) point out the following major reasons for mergers, acquisitions and partnerships amongst the various players in the telecom value chain:

1. Market Access: Enter into hitherto unexplored markets to acquire customers. This may be driven by saturation of the existing market/ geography or the desire to grow outside domestic markets.

Table 10.1   Category of Industry Segments in the Telecom Value Chain

Category	Description	Example
Network equipment manufacturers (NEMs)	Companies manufacturing hardware such as routers, switches, and multiplexers that are integral parts of the telecom network. Those that provide equipment to telecom service providers are referred to as carrier equipment manufacturers; and those that provide to enterprises are referred to as enterprise NEMs	Carrier NEMs: Alcatel-Lucent, Huawei ZTE, Nokia Siemens, Ericsson, Cisco, Nortel Enterprise NEMs: Avayya, Cisco, Siemens, 3Com, D-Link,
Handsets	Companies that manufacture and sell phones—fixed or mobile.	Fixed Phones: Bharti Teletech, National Panasonic Mobile Handsets: Nokia, Samsung, LG, Motorola, Sony Ericsson, Apple (iPhone), RIM (Blackberry), Google (Nexus One)
Network integration and management	Companies that provide network integration and management services	Wipro Infotech, HCL Comnet, HCL Info Systems, Datacraft
Network security	Companies that provide network security solutions	HCL Comnet, Wipro Infotech, Fortinet, Datacraft
Network storage	Companies that provide network-based data warehousing and storage solutions	IBM, EMC, HP, NetApps, Sun Microsystems
Audio and video conferencing solutions	Companies that provide audio and video conferencing platforms and services, mainly to enterprises	Polycom, Tandberg, Aethra
Telecom software	Companies that provide business support systems (BSS), operational support systems (OSS), embedded software for telecom equipment	Infosys, Tech Mahindra, TCS, Wipro, Sasken, Aricent

Test and measurement	Companies that provide test and measuring equipment for carriers and enterprises	Agilent Technologies, National Instruments, Spirent, Fluke Networks, Tektronix, Rohde & Schwarz
Turnkey contractors	Companies that do turnkey telecom projects	ORG, Telecom Consultants India Ltd., HFCL
Telecom cabling	Companies that provide structured cabling and associated solutions to either carriers or enterprises	Carrier Telecom Cables: D-Link, Systimax CommScope, AMP NetConnect Enterprise: Finolex, Sterlite, Birla Ericsson, Aksh Optic Fibre, Paramount Cables
Semiconductor design and manufacturing	Companies that design and manufacture semi-conductor chips that go into mobile handsets	Texas Instruments, Infineon, ST Ericsson, Broadcom
Semiconductor and embedded design services	Fabless design and development of embedded software for communication equipment and phones	Sasken, MindTree, Wipro
Tower companies	Build and operate towers for mobile base station transcievers	American Tower Corporation, GTL, Indus Towers, Reliance Infratel, Bharti Infratel

2. Economies of Scale and Scope: Acquire/merge to achieve complementary synergies and develop a more comprehensive product/service set. To achieve 'economies of scale and scope', firms acquire networks and customers from existing service providers for expanding their operations.

3. Controlling Technology: Enter into partnerships to control emerging technologies.

4. Technology Hedging: Enter into hitherto unknown technology areas due to the promise of futuristic technologies.

Box 10.1 provides some examples of telecom partnerships that have occurred with the above as the primary motivating factors (Sridhar, 9 September 2009).

Box 10.1   Telecom Partnerships to Ponder

With talks of Bharti-MTN Africa mega-merger valued at about US$ 23 billion on the horizon, the Indian telecom industry leads the rest of the sectors in merger and acquisition deals.

'Market access' is one of the primary motives for such partnerships. The mobile operators whose domestic market is either very small or getting saturated are looking for opportunities in emerging markets such as India to expand their presence. The acquisitions of stakes by Etisalat (UAE), Maxis Telecom (Malaysia), Bahrain Telecom (Bahrain), Telenor (Norway) respectively in the new entrants such as Swan Telecom, Aircel, S Tel and Unitech Wireless are examples of market access, with a combined value of more than US$2 billion. While the foreign operators are busy buying stakes in the new entrants, the large Indian incumbents are vying for opportunities outside the country. Thanks to the large number of licenses issued, the Indian mobile market has become hyper-competitive with thirteen to fourteen operators in each circle. This has forced larger incumbents such as Bharti Airtel and MTNL to look for opportunities outside the country to expand their footprint. The two US giants—American Tower Corp. and Crown Castle International, are slugging it out for the US$ 2 billion tower business, which is up for sale from Aircel, to gain entry into the high-growth Indian tower market.

After getting market access, firms enter into the second phase of building partnerships to achieve complementary synergies and develop a more comprehensive product/service set. To achieve 'economies of scale and scope', firms acquire networks and customers from existing service providers for expanding their operations. With the African market resembling parts of the Indian market, firms such as Bharti Airtel look for synergistic opportunities through merging operations in these two countries. Bharti, having acquired 1 million customers in Sri Lanka as the fifth mobile operator is eying a partnership with Millicom's Tigo Network, the third largest operator, for the same purpose. Idea Cellular's purchase of Spice brought together two regional players into a common fold, thus giving Idea a pan-Indian presence. The partnership between Indus Towers owned by Bharti-Vodafone-Idea consortium and Tata-Quippo are examples where the tower arms of the telecom companies are pooling their passive infrastructure and sharing it to lower their cost of operations. Bharti Airtel signed up with more than 100 mobile application/content providers including Infosys, Indiagames, and Hungama in its Service Delivery Platform (SDP) to achieve economies of scope in its service offerings. By bundling applications provided by their partners through the SDP, Bharti intends to gain over its competitors in the evolving VAS market. The platform also expands the scope of the content offerings of Bharti through its IPTV and DTH platforms. In the same vein, Reliance has entered into a strategic alliance with BBC World Service to offer 'live audio' thus expanding the scope of offerings in its R-World VAS platform.

The third major factor for the partnerships in telecom is for companies to have 'control over emerging technologies'. Many operators are tying up with companies that possess specific technology expertise. This is evident from the operators outsourcing their network management and IT services to network equipment vendors and large IT service vendors. While the Bharti-Ericsson-Nokia Siemens-IBM partnership broke the myth of outsourcing core telecom activities, it had a suite of followers in Vodafone Essar-Nokia Siemens-IBM, Idea Cellular-IBM, and Aircel-Wipro. Recently, Reliance and Alcatel Lucent joined hands to create a managed services company that would synergies expertise in network management, IT and technology. The MTNL which was given the 3G spectrum last December and is struggling with customer acquisition is planning to enter into a franchise partnership model which could bring in technology expertise and funding for its 3G network and service deployment. Tata Teleservices roped in DoCoMo, the leading mobile operator in Japan with proven expertise in mobile multimedia applications to give a fillip to its recently launched GSM services.

However, for the partnerships mentioned above to be successful, rightful policy directives are required. One of the outcomes of these partnerships, especially when done with the objective of achieving scale economies, is for consolidation of assets and operations. The recently stipulated three-year lock-in period for the new entrants and the excessive restrictions on intra-circle acquisitions (not more than 40 per cent market share for the merged entity and not more than 10 per cent cross-equity holding) may prevent consolidation in the market and impede market access by the foreign operators. While the hyper-competition results in a drop in prices temporarily, the excessive fragmentation of the market and the associated spectrum resources could be harmful for the industry in the long run. The government and the policymakers should temper the 'windfall gain' the operators, especially the new entrants, might make in such partnerships, through measures other than curtailing market forces. The partnership envisaged by the MTNL for its 3G services also might not find favour with the government, as the policy on Mobile Virtual Networking Operations, though recommended by the regulator in August 2008, is yet to see the light of day.

In telecom, regulation and policy are still to catch up with the lightning speed of evolution of technology and marketplaces.

As pointed in the Box 10.1 above, market access, economies of scale and scope, controlling technologies and technology hedging follow in that order. Table 10.2 lists the important telecom partnerships that have taken place around the world.

Table 10.2 Telecom Partnerships and their Primary Motivators

Year	Partnership	Reason
Market Access		
2000	Bharti acquiring JT Mobile	To get entry into Andhra Pradesh, Karnataka and Punjab service areas.
2001	Bharti acquiring a stake in Spice Telecom Kolkata	To enlarge its then fifteen service areas' footprint.
2002	Tata Teleservices acquiring a 56 per cent stake in Hughes Telecom (India) Ltd. (Hughes Tele.net) for Rs 3.64 billon	To gain access to the basic telecom market in Maharashtra.
2004	Idea Cellular acquiring Escotel for Rs 11.5 billion	To gain entry into Kerala, UP(W) and Haryana service areas.
2004	Bharti acquiring a 67.5 per cent stake in Hexacom India in Rajasthan for Rs 4.3 billion	To enter into the growing Rajasthan service area, thus getting the captive network and subscriber base of Hexacom.
2004	Telekom Malaysia acquiring a 47 per cent stake for US$ 390 million in Idea Cellular	To enter into the Indian market hedging risk in domestic markets.
2005	Birla AT&T acquiring Tata Cellular for US$ 44 billion; later buying RPG Cellcom in Madhya Pradesh	To enlarge footprint in areas where the acquiring company did not have networks and subscribers.
2005	Vodafone acquiring a 10 per cent stake in Bharti for Rs 65 billion	To enter into the lucrative Indian market.
2005	VSNL acquiring Teleglobe for US$ 239 million	To enter into wholesale voice, data, Internet services to about 240 countries in which Teleglobe had presence.
2005	Maxis Communications (Malaysia) acquiring a stake in Aircel for US$ 1 billon	To enter into the Indian market hedging risk in local markets.
2006	Tata group investing US$ 60 million for a 26 per cent stake in InfraCo in South Africa	To enter into emerging domestic and international markets in South Africa.

Year	Deal	Purpose
2004–6	Reliance acquiring FLAG, the submarine cable operator for US$ 211 million	To enter into international wholesale bandwidth business to provide connectivity to SEA-ME-WE countries.
2006	Bharti's stake worth $ 500 million in the SEA-ME-WE-4 project	To enter into international wholesale bandwidth business in many countries that are connected by the respective networks.
2007	Vodafone buying a 67 per percent stake in Hutchison Essar for US$ 10.9 billion	To enlarge its footprint in the Indian market; also to hedge against risks in its domestic and other international markets.
2006	Telekom Malaysia acquiring a 49 per cent stake in Spice Communications for US$ 178.8 million	To enter into the Indian market hedging risk in local markets.
	Virgin Mobile entering into a partnership with Tata Teleservices	To enter into the Indian market as an MVNO.
2007	Reliance Communications acquiring a stake in Yipes networks for US$ 300 million	To enter into the US data communications market.
2008	Tata Communications acquiring a 26 per cent stake in Neotel (South Africa)	To enter into a similar lucrative market in South Africa.
2008–9	Etisalat (UAE) acquiring a 45 per cent stake in Swan Telecom for US$ 900 million; Telenor (Norway) acquiring a 60 per cent stake in Unitech Wireless for Rs 61.2 billion; NTT DoCoMo acquiring a 26 per cent stake in Tata Teleservices for Rs 130.7 B; Bahrain Telecom (Bahrain) acquiring a stake in S Tel for US$ 1.25 billion; Sistema (Russia) increasing its stake in Shyam Telelink	To enter into the Indian market which opened up for more mobile operators in 2007.
2009	American Telecom Corporation buying XCEL Telecom	To enter into the lucrative mobile tower operations' business in India.

Economies of Scale and Scope

Year	Event	Purpose
2001	Bharti Airtel acquiring Skycell (Chennai)	To expand presence to leverage economies of scale.
2002	Tata group buying a 25 per cent stake in VSNL	To complement basic and cellular mobile business with VSNL's ILD and Internet services business.
2003	VSNL acquiring Dishnet DSL for US$ 130 million	To grow its Internet services presence and subscriber base.
2004	VSNL acquiring Tyco Global Network for US$ 136 million	To supplement its submarine cable business.
2007	Bharti Airtel increasing its stake in Network i2i	To enlarge its submarine cable and bandwidth operations.
2008–9	Bharti Airtel, Vodafone Essar and Idea Cellular partnering to form Quippo	To share and leverage the combined passive infrastructure in the telecom tower business.
2008	Idea Cellular acquiring a 49 per cent stake in Spice Communications for Rs 281 million	To expand its pan-India presence to include Karnataka and Punjab service areas thus augmenting its larger network effects.
2009	Bharti Airtel acquiring Warid Telecom (Bangladesh); Bidding to acquire the African operations of Kuwait-based Zain Telecom	To enter into a new market; also seeking synergies in operation.
2009	GTL Infrastructure buying Aircel's telecom tower business for US$ 1.8 billion	To enlarge its existing telecom tower business.
2009	American Tower Corporation acquiring Essar group's tower unit for Rs 20 billion	To expand its tower business in India and to grow larger to achieve economies of scale in operations.
2009	Reliance entering into strategic alliance with BBC World	To enlarge the scope of offerings including media and TV on mobiles in its R-World VAS platform.
2009	Bharti Airtel entering into alliances with more than 100 mobile application providers such as Infosys, Hungama and Indiagames	To enlarge the scope of its mobile VAS business.
2009	Aircel entering into a partnership with Infosys to provide mobile applications on it Flypp platform	To enlarge the scope of VAS offerings.

Technology Control

2003	Bharti outsourcing network deployment and management to Ericsson and Nokia Siemens for US$ 724 million	To leverage on the expertise of the NEMs to provide state-of-the-art services by transferring technology obsolescence and infrastructure upgrading risk to vendor.
2003	Bharti outsourcing IT management to IBM Ericsson and Nokia Siemens for US$ 750 million	To leverage on the expertise of the vendor in developing IT services to meet the needs of the future.
2004–9	Vodafone Essar outsourcing to Nokia Siemens and IBM; Idea Cellular to IBM; Aircel to Wipro	To leverage on the expertise of the vendor in developing telecom IT services to meet the needs of the future.
2008	Bharti Airtel entering into a strategic alliance with Verisign	To bundle security products and services in the ever evolving insecure telecom networks.
2009	Reliance Communications and Alcatel Lucent creating a separate managed services company	To leverage on the expertise of each other and to control the technology and services space.
2009	Tata Teleservices roped in DoCoMo (Japan)	To leverage DoCoMo's expertise in multimedia applications and services.
2009	Bharti Airtel entering into a strategic alliance with Cisco	To leverage Cisco's enterprise solutions including Telepresence to provide advanced business and enterprise services.

Technology Hedging

2005	Reliance acquiring a 51 per cent stake in movie production company Adlabs	To enter into content provisioning such as video-on-demand, mobile TV, and digital terrestrial broadcasting.
2008–9	Unitech (real-estate) and Datacom (consumer electronics) getting 2G mobile licence	To enter into the lucrative telecom services business.
2009	Videocon (consumer electronics such as TV, refrigerators) entering into mobile handset manufacturing	To enter into the growing telecom handset business.

Source: Tele.net, 2010; Sridhar, 22 December 2003; Sridhar, 12 September 2006; Sridhar, 26 March 2008.

The number of partnerships in the recent years has increased with economies of scale and scope and technology control being the primary motivations. Though technology hedging has not seen much activity, it is expected that it will increase in the years to come. Box 10.2 describes the state of the telecom industry and what holds for the future (Sridhar and Venkatesh, 2010).

Box 10.2   What is in Store for the Mobile Telecom Industry?

Global capex spend on telecom declined by almost 6 per cent in 2009 and is expected to hit rock bottom in 2010. The service providers in the advanced markets, hit by a drop in consumer spending, have cut down or postponed their decisions on networks and equipment. In effect, there is a conscious effort by all the stakeholders in the telecom supply chain to be extremely cautious about their cost structure and the need for extracting more value out of every investment made.

However, a glimmer of hope still persists in the emerging telecom markets in China and India which continue to witness hyper growth. The mobile segment in India witnessed an addition of about 185 million subscribers in 2009. The addition of three to five operators in each circle has intensified competition. This shift in growth and the associated telecom equipment spending has drawn the attention of the global players to reach and serve these growing geographies. However, the associated challenge in this market for the global telecom gear-makers is to reduce costs to be competitive, especially against the low-cost Chinese manufacturers, though security concerns regarding Chinese vendors still linger. The consolidation in the network equipment space continues. The purchase of Nortel's CDMA and LTE technologies by Ericsson, has forced the three European and one North American vendor in the network equipment space to compete rigorously, especially on price, with a large Chinese counterpart. This makes the case for the promotion of a domestic equipment vendor in India who can possibly satisfy the specific needs of the market.

There is intense growth in the Smartphone segment. Worldwide shipments of Smartphones are expected to surpass 350 million units, growing at a compound annual growth rate of 20.9 per cent during 2009–2013. The share of Apple's iPhone in the Smartphone market segment increased to about 19 per cent, while Google's Nexus One launched for the first time in the US directly into the retail market also witnessed good sales. The US$ 1.2 billion acquisition of Palm by HP has thrown in one more serious contender. However, we have also been witnessing commoditization of the Smartphone components such as audio/video players, browsers, codecs and even chipsets. Taiwan-based MediaTek has become a serious contender in the Smartphone segment, especially in India with the launch of low-cost feature-rich Smartphones.

There is also hyperactivity in the Smartphone platforms' market. There are a host of application development platforms today such as the Android promoted by the Open Handset Alliance and Google, Apple's iPhone OS 4, Nokia's Symbian, webOS of Palm/HP, and Blackberry of RIM amongst others. Nokia is trying to resurrect itself in the high-end Smartphone market by adopting Intel processor-based Meego platforms, thus enabling PC and mobile convergence. While this provides the content and application developers a huge opportunity to build applications and solutions that can be cross-sold across a number of platforms, the clear winner is yet to emerge.

The growth in Smartphones has contributed substantially to the increasing adoption of mobile broadband. It is expected that one in every three mobile subscribers will adopt mobile broadband, resulting in about 1 billion mobile broadband users by the year 2012. However, the battle between content/application providers and the telecom companies continues. The mobile operators, especially in the US, have started putting restrictions on the applications that can run on their 3G networks, citing huge downloads clogging the networks as the reason. The proponents of 'net neutrality' want prohibition against such blockages of content, applications and mobile Internet access. In a much bandwidth-constrained 3G services market in India where each operator gets only 2 × 5 MHz, the walled-garden approach of the operators is likely to continue to the detriment of subscriber interests and preferences.

The long-awaited 3G auction is over in India, with operators committing about Rs 50,000 crores for acquiring minimal spectrum. With the auctioning of airwaves for BWA spectrum and the emergence of new technologies such as TD-LTE as a competitor to WiMax, the battle for mobile broadband has intensified in India. However, it is to be seen whether India will witness the 'winner's curse' of operators paying much above the expected valuation for the 3G and BWA spectrum in uncertain demand conditions, much similar to what was witnessed in India in 1995 and later in Europe in early 2000. The TRAI's recommendation of fixing the price of 2G spectrum to 3G spectrum auction price, that too post the auction, has thrown a spanner into the business plans of the incumbents. A long litigation battle is awaited.

The mobile telecom sector has been a golden goose for the Indian government and has provided the much needed triggers for the economic development of the country. We can only hope that the concerned stakeholders act responsibly in providing a stable policy and development environment for the sector to reach much greater heights to maximize social welfare—that is to maximize the sum of utilities of the consumers, operators and the government together and not pit one against the other.

Regulatory Bottlenecks for Partnerships

Regulations play a big role in facilitating partnerships, alliances, mergers and acquisitions. However, there are many regulatory bottlenecks in India for the partnerships amongst firms to succeed and these are listed below (Prasad and Sridhar, 14 August 2008):

1. The minimum number of mobile access providers required in each licence area is set at four. This will prevent mergers and acquisitions in several circles where efficient scales are achieved with two to three operators. To have a minimum level of competition as well as industry efficiency, Jagdish Seth's classic 'rule of three' should be applied, and the minimum number of operators is set to three. Alternatively, different thresholds could apply to different categories of circles with a minimum of three.
2. The maximum market share of the merged entity has been set at 40 per cent. In some service areas, efficient scales can be present only above this limit which will prevent industry efficiency and prevent mergers and acquisitions.
3. The maximum cross-holding by an entity in a service area can be 10 per cent. In the case when the combined market share of the cross-held entities is less than the stipulated cap, the law is unnecessarily restrictive. A holding of 10 per cent or less in most cases is in the nature of a financial investment, and therefore need not be regulated unless the market share cap is violated by the merged entity. In many countries the cap on the market share is the primary instrument used to control market power and where possible we need to adopt the same approach. The European Union regulation even says that a higher market share does not necessarily lead to higher market power and sustainable price increases.
4. A cap of 74 per cent on FDI prevents a foreign operator from seting up operations directly in India.

The Competition Commission of India (CCI) is supposed to define and measure 'significant market power' and prescribes directives accordingly. However, the CCI has not been active in the telecom partnerships. The regulatiory clauses should be carefully reviewed by a competent authority to foster growth of the sector and promote industry efficiency.

Convergence and Its Implications

Convergence in communications is being discussed for a long time since the enactment of the Telecommunications Act of 1996 in the US. However, convergence in all aspects of telecom is yet to happen. The industry and regulation as discussed in the previous chapters are structured along different compartmentalized segments.

Convergence can be explained as the blurring of dividing lines among traditionally distinct products and services, technologies, industries, and markets. The Internet is a powerful medium, which has embraced convergence quite effectively. Internet provides (i) *communication* services such as email and Internet telephony (ii) *broadcasting* services such as web-casting and electronic community services (iii) *publication* services such as publishing electronic journals, e-books and electronic manuals on the World Wide Web and (iv) *electronic commerce* and electronic business services. The Internet with its wide offerings of services is making inroads into traditionally distinct areas such as telephone and cable TV.

Technology convergence has enabled broadband Internet connectivity and video-on-demand (IPTV) to be provided through normal telephone lines; video sharing over mobile phones; and Internet connectivity through cable TV networks and DTH satellite antennas. Products and services are being bundled in today's convergent market. Most of the telecom companies have bundled telephone connections with Internet connectivity—the LCO has bundled cable TV with Internet service and mobile service providers have started offering Internet access and broadband content along with voice services. The industries and market have responded to the convergence in technologies, products and services. Figure 10.1 explains how convergence is taking place across different areas of telecom (Sridhar, 2000).

In the innermost circle is the consumer who uses different devices such as telephone, TV, computer or a cell phone. The consumer uses these devices to obtain different services such as plain old telephone service, cable TV service, Internet access or cell phone services. These services are being provided through appropriate networks by service providers such as telephone companies, cable TV operators, ISPs, or cellular operators. There is convergence in every entity starting from end-user devices all the way to the service providers.

Today, technologies have been developed so that voice, data, and video messages can be transported through a single physical telecom

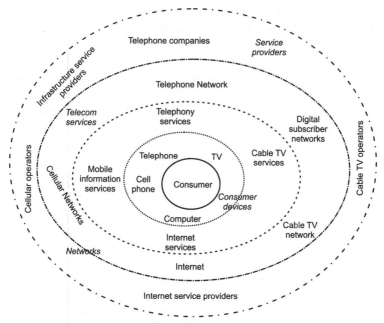

Figure 10.1 Communication Convergence

infrastructure. Internet telephony provides telephony services at about a fraction of the cost of today's regular telephone calls, especially for STD and ISD services, though the present regulations in India are restrictive as discussed in an earlier chapter. Computer and telephone have converged to provide integrated service in this case.

Cable networks can be used not only for video broadcasting, but also for Internet connectivity using cable modem technology. Here we see different networks such as the cable network and the Internet seamlessly integrated to provide bundling of services to the consumer. This convergence in networks has led to convergence in the telecom industry with alliances between cable operators and ISPs. The disastrous merger between America On Line (AOL), the largest ISP in the world and Time Warner, the largest media company is an interesting case in this regard. AOL, which provided Internet services, foresaw content created by Time Warner Communications, such as movies, to be delivered through their network. The AOL network could be used by the cable TV network of Time Warner to provide Internet access.

Access technologies such as asymmetric DSL (ADSL), offered by the BTOs seamlessly integrate voice, data and video (for example, IPTV) services. By and large, Internet access has been a wire-line-based activity. However, with the availability of 3G and BWA, mobile Internet has come of age and is likely to be the preferred option in the landline-constrained environment in the country.

With such convergence of information and communication technologies, it is now possible for service providers to bundle different services so that consumers can purchase services from a single service provider for all their requirements. It is also possible for consumers to access different services from anywhere through a single intelligent device. The day is not far off when we can all instruct our intelligent home washing machines connected to the Internet, to start operation right from our desktop office computer.

However, the regulations and polices have traditionally been lagging. The Convergence Bill created in 2000 recommended the creation of a 'composite telecom license' and the setting up of a super regulator called the 'Communication Commission of India'. However, the Bill was never tabled in Parliament. Though unification has been achieved to some extent between basic telecom service and cellular mobile service through the implementation of the UASL, removal of restriction on the Internet telephony service to be provided by the ISPs, unbundling of the local loop, and other such regulatory interventions are required to bring in true communication convergence.

It has been argued by many researchers that there is positive association between telecom penetration and economic development. This association is augmented by convergence of products and services. Though the TRAI has been instrumental in implementing CAS and for the emergence of DTH in the country, a separate Broadcast Regulatory Authority under the Ministry of Information and Broadcasting is often mooted. However, more regulators translate to more control and micromanagement (Sridhar, 2007).

Telecom growth and convergence very much depend on stable and forward-looking regulatory policies which are implemented in tune with technology deployment. Our research shows that one of the primary factors that is driving the exponential growth of mobile services in India is competition. The Telecom Act of 1996 that introduced competition in the US telecom market spawned the development of new technologies such as digital subscriber loop and cable modem by erstwhile monopoly operators in their fight for survival. British Telecom, the government operator in the UK is taking a proactive step in the deployment of next-generation networks to lead the pack. With policies directed towards healthy competition in the marketplace, we can definitely hope for another revolution in the country—growth of broadband and convergence.

Glossary

1G   First-generation—The analogue mobile telecommunication systems for voice services. Advanced Mobile Phone System (AMPS) was one of the widely used analog systems in the 1980s.

2G   Second-generation—Digital mobile telecommunication systems that provides basic voice and minimum speed data connectivity. It uses circuit switching for voice and packet-switching for data communication. GSM is the widely used standard in 2G while CDMA is a proprietary standard developed by Qualcomm, USA.

3G   Third-generation—Digital mobile telecommunication systems that provide high-speed data connectivity and enable video calling, mobile TV and other multimedia services apart from supporting high quality voice service. The specifications for 3G were set up by the International Mobile Telecommunications-2000 (IMT-2000) forum of the ITU. The 3GPP provides standardization support for 3G services. To meet the IMT-2000 standards, a system is required to provide peak data rates of at least 200 kbit/s. Recent 3G releases, often denoted 3.5G and 3.75G, also provide mobile broadband access of several Mbit/s to smartphones and mobile modems in laptop computers. WCDMA, CDMA-EVDO, HSPA, HSDPA, HSUPA are examples of technologies that come under the umbrella of 3G services.

3GPP   Third-generation Partnership Project—The industry group that oversees the creation of industry standards for the 3rd generation of mobile wireless communication systems. The key members of the 3GPP include standards agencies from Japan, Europe, Korea, China, and the United States. More information about 3GPP can be found at www.3GPP.org

4G   Fourth-generation—Mobile telecommunication systems offering significantly high speed data and multimedia services (typically 100 Mbps or more) and interoperable with previous generation systems as well as various communication technologies.

ADC   Access Deficit Charges—Charge levied on non-BTOs to be transferred to BTOs to enable them to cover the difference between cost and revenue of laying down often expensive wired local loops.

ADSL   Asymmetric DSL—Communication system that provides asymmetric capacities across uplink (from CPE to network) and downlink (from network to CPE) for the transfer of digital information across the DSL. Normally provided by the BTO, the data transmission rate varies depending on distance of CPE from the network node, line distortion, and settings.

AGR   Adjusted Gross Revenue—The gross revenue accruing to the telecom licensees by way of operations of the cellular mobile service and also included revenue on account of value-added services, supplementary services, and sale of handsets plus revenue accruing through re-sellers, and franchisees.

APNIC   Asia Pacific Network Information Centre—An open, membership-based, not-for-profit organization providing Internet addressing services to the Asia Pacific. More information can be found at www.apnic.net

ARPU   Average Revenue Per User—Revenue accrued to the telecom operator per subscriber per month.

ATCA   Advanced Telecommunication Computing Architecture— The standard developed by PCI Industrial Computer Manufacturers Group with specifications related to high speed interconnect technologies, next generation processors, and improved reliability, availability, and serviceability of systems.

ATM   Asynchronous Transfer Mode—Data link protocol used in high-speed networks. ATM networks are typically deployed in the backbone of packet switched networks. Optic fibre communication links typically connect ATM switches at different locations. Local ATM switches enable high-speed data communication within and across enterprise LANs.

Auction   Method of allocating scarce resource such as spectrum in telecom, especially for mobile services. The potential buyers bid for the block of spectrum sold. The auction can be a single stage or multi-stage, open or closed type. Spectrum for all LSAs can be simultaneously or sequentially auctioned. Typically these auctions are strictly ascending where the bid price increases in each stage and the highest bidder(s) is designated as the winner. There can be reverse auctions as well, especially for distributing subsidy for telecom services in which case the auction is a descending auction and the lowest bidder is designated as the winner.

AUSPI   Association of Unified telecom Service Providers of India (AUSPI)—The representative industry body of UASPs providing CDMA & GSM Mobile Services, Fixed Line Services as well as Value Added Services throughout the length and breadth of the country. More information can be found at www.auspi.in

Beauty   Concept developed by the famous economist John
Contest   Maynard Keynes and is practiced in the allocation of licenses/spectrum in telecom. The potential buyers of the licenses are required to submit values and associated proofs for various parameters (viz. telecom experience, rentals they will charge, investment they will make). The weightages for each parameter might vary. The weighted sum of values assigned for the buyers is used as a measure to allot licenses/spectrum. This method was used in countries such as Finland for allocating 2G and 3G mobile licenses and associated spectrum.

BFSI   Banking, Financial Services, and Insurance—The sector consisting of banks and financial institutions. The telecom and IT investment in this sector is substantial.

BSC   Base Station Controller—Automatic coordinator (controller) that permits one or more BTSs in a wireless network to communicate with a MSC.

BSNL   Bharat Sanchar Nigam Limited—Created in 2000 by the Government of India to provide telecom access services across the country except in Delhi and Mumbai service areas. Also provides NLD services. More information can be found at www.bsnl.co.in

BSS   Business Support System—The IT enabled business systems dealing with customers, supporting processes such as taking orders, processing bills, and collecting payments, typically of telecommunication carriers.

BTO   Basic Telecom Operator—Telecommunication company that has received the licence to provide land line or fixed wireless based voice and data communication access services.

BTS   Base Transceiver Station—The Radio Access Network node of the cellular mobile network that communicates with the mobile user terminals or handsets. BTS is connected to Base Station Controller (BSC). BTS with its antenna, transmitter and receivers are normally mounted on cellular towers to provide limited geographical coverage.

BWA   Broadband Wireless Access—Provisioning of high-speed data access and connectivity through Radio Frequencies.

CAS   Conditional Access System—System for providing selective access to subscribed channels on the cable TV network.

CDMA   Code Division Multiple Access—The multiple access scheme developed by Qualcomm Inc., USA to support multiple simultaneous communications over the wide band frequency by adding a unique code for each data signal that is being sent to and from each of the radio transceivers. CDMA development group is a consortium of companies that promote the evolution of CDMA. More information can be found at www.cdg.org

CDMA-EV-DO   CDMA–Evolution Data Optimized—A 3G digital service provided by CDMA cellular carriers. Part of the CDMA2000 standards, EV-DO system uses one or more 1.23 MHz radio channel bandwidth as the existing CDMA 2000 system that provides for multiple voice channels and medium rate data services.

C-DOT   Centre for Development of Telematics—Created in

	August 1984 as government-owned company with the specific intention of indigenizing digital switching technology to meet India's unique requirements. C-DOT was established as an independent society to help develop a series of digital switching products to meet Indian requirements. More information can be found at www.cdot.com
CLS	Cable Landing Station—The place where the submarine optic fibre cable makes the landfall. It is also the point at which the International Gateway is set up to interconnect with domestic networks.
Cellular networks	Wireless communication networks that consists of many cells containing BTSs, covering the licensed service area of an operator. The non-adjacent cells typically re-use frequencies in order to optimize the use of available spectrum resources.
Circuit-switching	Method of connecting a communicating pair of nodes in a network using dedicated channels. Normally used in voice services.
CMTS	Cellular Mobile Telecommunication Services—Communication services that is provided by dividing a specific geographic area in to 'cells'. Each cell typically extends fro few 100 metres to about 4–5 kms. Each cell is serviced by a BTS.
COAI	Cellular Operators Association of India—Constituted in 1995 as a registered, non-governmental society, the Association is dedicated to the advancement of modern communication through the establishment of a world-class cellular infrastructure and to delivering the benefits of affordable mobile communication services to the people of India. More information can be found at www.coai.com
CPE	Customer Premise Equipment—Devices such as telephone, computer, modem at the customer's residence or office that is connected to the network.
CSCs	Community Service Centres—Locations that provide high quality and cost-effective video, voice and data content and services, in various areas of e-governance as well as other private services, funding for which is provided by the state/central governments.

CUG	Closed User Group—Group of users belonging to an organized entity; often having common communication requirements
DAMA	Demand Assigned Multiple Access—The scheme used in satellite channels allocation. Satellite channels are grouped together as a bulk asset, and DAMA assigns users variable time slots matching user information transmission requirements.
DECT	Digital Enhanced Cordless Telephony—Digital communication standard originated in Europe, primarily used in cordless phone systems.
DLC	Domestic Private Leased Circuit—The private (not shared) transmission/ communication link set up between customer premises across locations (typically across LDCAs) within the country, service for which is typically provided by the NLDOs.
DLC	Digital Loop Carrier—The device/system that are typically deployed near the customer premise. Carriers run fibre cabling from their central offices to DLCs from where the twisted pair copper cable connects the end user CPEs.
Domain Name	Domain refers to a group of networks that are under the administrative control of a single entity, such as a company or a government agency. Domain name is given by ICANN and ICANN accredited agencies.
DoS	Department of Space—The Space Commission formulates the policies and oversees the implementation of the Indian space programme to promote the development and application of space science and technology for the socio-economic benefit of the country. DOS implements these programmes through, mainly Indian Space Research Organisation (ISRO), Physical Research Laboratory (PRL), National Atmospheric Research Laboratory (NARL), North Eastern-space Applications Centre (NE-SAC) and Semi-conductor Laboratory (SCL).
DoT	Department of Telecommunications—Department under the Ministry of Communications and IT of the GoI that oversees the telecommunications sector of the country including policymaking, spectrum allocation and licensing. More information can be found at www.dot.gov.in

DSA   Dynamic Spectrum Access—Method of allocating either licensed or unlicensed spectrum dynamically using co-operative or opportunistic manner. DSA is a feature in Cognitive Radio networks.

DSL   Digital Subscriber Loop—Transmission/service that provides digital transmission over copper wires that are traditionally used for voice telephony. The service is accessed using a modem and is normally provided by the BTOs.

DTH   Direct-to-home—Television broadcasting service that uses Ku and associated high frequency radio frequencies to transmit video programmes from geo-stationary satellites to small antennas mounted on the roof top of households.

DTMF   Dual Tone Multi Frequency—The telecommunication signalling over analog telephone lines in the voice-frequency band between telephone handsets and other communications devices and the switching centre. The version of DTMF that is used in push-button telephones for tone dialling is known as Touch-Tone. Variety of Interactive Voice Response services are provided using DTMF.

EDGE   Enhanced Data rates for GSM Evolution—High speed packet data service offered by mobile telcos over GSM networks. This is a superior data offering compared to GPRS.

EMS   Electronic Manufacturing Services—Design, test, manufacture, distribute, and provide return/repair services for electronic components and assemblies for OEMs. It is also referred to as Electronic Contract Manufacturing.

EPABX   Electronic Private Access Branch Exchange—A small exchange normally hosted in the enterprise to cater to its communication needs. Typically it enables voice communication across the user terminals connected to it. The exchange is typically connected to public network using leased lines.

ERNET   Engineering Research NETwork—Created by the Department of Electronics, Ministry of IT, GoI, with the funding from United Nations Development Programme. The objective was to create expertise R&D and education in the country in the area of networking and Internet in the country. ERNET is the largest nationwide terrestrial and satellite network in India with point of presence located at the premiere educational and research institutions in major cities of the country. More information can be found at www.eis.ernet.in

ERP   Enterprise Resource Planning—IT system that enables different organizational process including order processing, inventory management, human resources management.

Ethernet   Developed by Robert Metcalf at Xerox, Palo Alto, USA, in 1976, this Local Area Network (LAN) standard has become ubiquitous for connecting devices within home and office buildings. Later adopted by IEEE as 802.3 Code Division Multiple Access (CDMA)/Collision Detection (CD) standard, Ethernet has evolved in transmission speeds from 10 Mbps in the 1970s to 100 Mbps in the 1980s and now 1000 Mbps (Gigabit Ethernet). Carrier grade Ethernet are also being deployed by the telcos using optic fibre or fibre-coaxial cable to provide wired broadband connectivity.

Extended   Radio frequencies in the range of 3.4–6.425 GHz
C Band   normally used for VSAT data communications and TV broadcasting.

FCC   Federal Communications Commission—The independent regulatory and licensing authority for the telecommunications and broadcasting sector in the USA, set up by the Communications Act of 1934.

FCFC   First Come First Serve—Method of allocating telecom licence/spectrum based on who applied/ paid the fee first.

FDMA   Frequency Division Multiple Access—Process of allowing mobile radios to share radio frequency allocation by dividing up that allocation into separate radio channels where each radio device can communicate on a single radio channel during communication.

FM   Public over the air radio broadcasting using Frequency
Broadcasting   Modulation (FM). The FM Radio can be received through an antenna normally provided in devices such as mobile handsets, and car radio units. FM broadcasting is typically provided over Very High Frequency (VHF) Signals in the 30–300 MHz range. The propagation is generally limited

	compared to Amplitude Modualtion (AM) broadcasting to a range of about 50–100 miles.
FTA	Free To Air—The television channels that need to be provided by the LCO/ MSO over the air/ cable, thus not requiring the installation of cable STB. The LCO/ MSO needs to provide the minimum number of FTA channels especially in CAS notified areas at designated tariff. This is to provide option to cable TV subscribers to choose between pay channels and FTA channels. Typically the FTA channels cover news, education and health.
GFN	Group Forming Networks—These are networks of groups having a common interest.
GoI	Government of India
GPRS	General Packet Radio Service—A basic packet data always on-line service provided over GSM networks.
GSM	Groupe Speciale Mobile (Global Systems for Mobile)—A public all-digital cellular network standard using the transmission band around 900 MHz, 1800 MHz, and 1900 MHz developed by the European Telecommunication Standards Institute (ETSI). A GSM network can provide, besides telephony services, Short Messaging Services (SMS) and data communication, in circuit- and/or packet mode. GSM Association represents the corresponding interest group. More information is available at www.gsm.org
H.323	An ITU based standard for sending voice though IP.
HHI	Herfindahl-Hirschman Index—Measure of the size of firms in relation to the industry and an indicator of the amount of competition among them. It is calculated as $\sum s_i^2$ where s_i is the share of the firm 1 (for example, subscribers) in the market. A value of 1 indicates monopoly, while 0.18 indicates perfect competition.
HITS	Head-end in the Sky—A system in which the encrypted television content in digital form is broadcast to LCOs so that the same can be retransmitted to residences by the LCO through its cable TV network. The HITS operator can broadcast either its own content or content syndicated from other broadcasters or simply provide passive platform for other MSOs or broadcasters.

HSDPA	High Speed Downlink Packet Access—High speed 3G digital data services based on WCDMA with a downlink speed of 14.4 Mbps and uplink speed of 400 Kbps.
HSPA	A family of high-speed 3G digital data services based on WCDMA, with a downlink speed of 42 Mbps and uplink speed of 5.8 Mbps.
HSUPA	High Speed Uplink Packet Access—High speed 3G digital data services based on WCDMA with a downlink speed of 14.4 Mbps and uplink speed of 5.7 Mbps.
ICANN	Internet Corporation for Assigned Names and Numbers—Formed in 1998, as a not-for-profit public-benefit corporation with participants from all over the world dedicated to keeping the Internet secure, stable and interoperable. It promotes competition and develops policy on the Internet's unique identifiers. ICANN does not control content on the Internet. More information can be found at www.icann.org
IEEE	Institute of Electrical and Electronics Engineers—The world's largest professional association dedicated to advancing technological innovation and excellence for the benefit of humanity. More information can be found at www.ieee.org
IETF	Internet Engineering Task Force—A large open international community of network designers, operators, vendors, and researchers concerned with the evolution of the Internet architecture and the smooth operation of the Internet. More information can be found at www.ietf.org
ILD	International Long Distance—Voice communication services across two countries, typically provided by ILD Operators (ILDOs).
INSAT	Indian National SATellite—A system of satellites that are placed in Geo-stationary orbits in Asia-Pacific region by ISRO. More information is available at http://www.isro.org/satellites/geostationary.aspx
Internet	The public network started by the Department of Defence in the USA, later adopted by educational institutions and corporates, providing data communication across user devises connected to it using TCP/IP.
Internet Telephony	Telephony service over the public Internet.

Intranet	Private network catering to the communication requirements of organizations using TCP/IP. Typically these private networks are connected to the public Internet using communication links.
IP	Internet Protocol—The default routing protocol of the Internet that operates at the network layer. Provides addressing and packet routing support. The IP address version 4 that is commonly used on the Internet thus far providing 32-bit address support is now being replaced by the 48-bit IP version 6. The IP uses connection-less datagram routing to transport packets hop-by-hop from the origin to the destination in the network.
IP TV	Internet Protocol Television—Television content streamed as packetized transmission over IP networks to customers who can access the same through their broadband connection on their PCs/Laptops. This service is typically provided by the BTOs who provide wired DSL based broadband connectivity to retail households.
IPLC	International Private Leased Circuit—The private (not shared) transmission/communication link set up between customer premises in two countries, service for which is typically provided by the ILDOs.
IPR	Intellectual Property Rights—A number of distinct types of creations of the mind for which a set of exclusive rights are recognised. Common types of IPRs include copyrights, trademarks, and patents.
ISP	Internet Service Provider—The firm that has got the appropriate licence to provide public as well as customized private Internet service.
ISPAI	Internet Service Providers Association of India—The association of ISPs in the country to promote Internet and associated services. More information can be found at www.ispai.in
ISRO	Indian Space Research Organization—The organization set by the GoI to develop space technology and its application to various national tasks. The Antrix Corporation, established in 1992 as a government-owned company, markets the space products and services developed by ISRO.

ITI	Indian Telephone Industries—The oldest public sector enterprise and the largest telecom manufacturing monolith in the country, conceptualized and started in the year 1948 as the first public sector company after independence.
ITSP	Internet Telephony Service Provider—An ISP licensed to provide Internet Telephony Service.
ITU	International Telecommunications Union—ITU is the United Nations specialized agency for information and communication technologies (ICTs). ITU allocates global radio spectrum and satellite orbits, develop the technical standards that ensure networks and technologies seamlessly interconnect, and strive to improve access to ICTs to underserved communities worldwide. More information can be found at www.itu.int
IUC	Interconnection Usage Charges—Charges that are levied by one telecom operator to another for interconnecting their networks.
Ku Band	Portion of the electromagnetic spectrum in the range of 10–14 GHz normally used for DTH services. The high frequencies enable the use of small dish antennas for capture.
LAN	Local Area Network—A communication network (typically for data transmission) that interconnects devices such as computers and printers within buildings.
LCO	Local Cable Operator—Registered service provider who provides local cable TV service.
LDCA	Long Distance Charging Area—Telecom networks are divided in to a number of LDCAs. Each LDCA contains one or more SDCAs. Each LDCA has a Long Distance Charging Centre (LDCC) which is a Trunk Automatic Exchange (TAC).
LoS	Line of Sight—Arrangement where the transmitter and receiver of information through wireless Radio Frequencies need to see each other.
LLU	Local Loop Unbundling—Leasing or renting of the wired local loop partly or wholly by the incumbent BTOs to new competing access providers or ISPs.
Local Loop	The copper pair wires (wired) or radio frequencies (wireless) that provide connectivity between the user

terminal/ CPE to the communications network, often referred to as the last mile.

LSA — Licensed Service Area—The geographical jurisdiction allotted by the licensor (the government) to a licensee (the operator) for providing communication service.

LTE — Long Term Evolution—The next-generation 4G technology for both GSM and CDMA cellular carriers. Approved in 2008 by 3GPP with download speeds up to 173 Mbps.

MAN — Metropolitan Area Network—A communication network that spans across geographical locations within a city/ town. MAN normally connects many different LANs within the city/town.

MNO — Mobile Network Operators—A licensed telecom operator who has spectrum and other associated facilities to provide mobile communication services.

MNP — Mobile Number Portability—The facility to keep the same mobile number while switching (porting) to a different operator/ geography/ service.

Mobile TV — Systems and technologies that provides broadcast TV over mobile networks to smaller handheld user terminals.

Modem — Device that connects the user terminal/device to a communication network, often providing some transformation to the user information such as converting the information in to appropriate digital/analogue signals.

MSC — Mobile Switching Centre—An interconnection system that is used to connect communication paths within mobile communication networks (cellular and 3G.)

MSO — Multi Service Operator—Registered firm providing bundling of content to LCOs, often including their own TV channels. In CAS notified areas, MSOs provide STBs and associated smart cards to end users through their LCO partners. Typically MSOs are larger and often cover a wide geographical region compared to LCOs; have optic fibre-based networks in cities. Most of the MSOs are ISPs providing Internet Services as well.

MTNL — Mahanagar Telephone Nigam Limited—Created in 1986 by the Government of India to provide telecom access services in Delhi and Mumbai. More information can be found at www.mtnl.co.in

Multimedia   A combination of voice, data, image and video, embedded in applications and services.

MVNO   Mobile Virtual Network Operator—A licensed/ registered telecom operator who normally leases/rents spectrum and other associated facilities to provide mobile communication services.

NAP   Network Access Point—The point of interconnect where the ISPs are connected to peer networks.

NEM   Network Equipment Manufacturer—Firm that produces communication network equipment such as switches, routers and other associated hardware/software.

NIXI   National Internet Exchange of India—The neutral meeting point of the ISPs in India. Its main purpose is to facilitate exchange of domestic Internet traffic between the peering ISP members. More information can be found at www.nixi.in

NLD   National Long Distance—Voice communication services that typically occur across LSAs, typically provided by NLD Operators (NLDOs).

NOFA   National Optic Fibre Agency—A 100 per cent central government-owned entity to plan, install, operate, and maintain shared fibre network in identified cities and provide means to allow any service provider to use the network for giving broadband connections using any technology in the last mile.

NTP 1994   National Telecom Policy 1994—The policy created by the Government of India in 1994 as a part of the liberalization process to introduce reforms in telecommunications services thus far provided by the government.

NTP 1999   New Telecom Policy 1999—The revised tele-communications policy announced in 1999 that took stock of the effects of NTP 1994 with suggestive additional reforms.

OEM   Original Equipment Manufacturer—A firm that builds a communication equipment such as handset or switch, though bringing different hardware and software components from contract suppliers, and taking ultimate ownership of the product.

OSS	Operational Support System—The IT products/ services that support processes in communication networks such as maintaining network inventory, provisioning services, configuring network components, and managing faults.
Packet-switching	Method of transmitting discrete units of information called 'packets' using a store-and-forward mechanism between two communicating node pairs in a network.
PCO	Public Call Office—Public manned booth where users can avail telephony services using manual/automated phone.
PDA	Personal Digital Assistant—Device that enables users to run applications such as enterprise tools, office automation software, apart from providing voice telephony and Internet connectivity.
PLMN	Public Land Mobile Network—The mobile telecommunication network that provides voice/data services.
PoI	Point of Interconnect—A PoP at which the networks of different operators interconnect.
PoP	Point of Presence—The point at which telephone and data calls are handed off between access providers and NLD operators or ISPs.
PSTN	Public Switched Telephone Network—The telecommunication network typically with wire-line access, that provides voice/data services.
PSU	Public Sector Unit—A government-owned corporation, state-owned company, state-owned entity, state enterprise, publicly-owned corporation, government business enterprise, or parastatal, either partly or wholly owned by the GoI, is a legal entity created by a government to undertake commercial activities on behalf of an owner government. Their legal status varies from being a part of government into stock companies with a state as a regular stockholder. The defining characteristics are that they have a distinct legal form and they are established to operate in commercial affairs. While they may also have public policy objectives, GOCs/ PSUs are differentiated from other forms of government agencies or state entities in that they do not have to pursue purely non-financial objectives. They are responsible for satisfying the shareholders' interests and hence are expected to produce

reasonable return on their investment through dividends. While BSNL and ITI are purely government owned, MTNL is partly government owned as (about 56 per cent) and the rest of its equity is offered to other shareholders including institutional investors and common public and is listed in the Indian stock exchanges.

PTT   Post, Telephone and Telegraph—A term used for a government agency in many countries that supplies and maintains the infrastructure and provides basic telecommunication services.

QoS   Quality of Service—Measure of quality of telecom service.

RoW   Right of Way—Legally granted access to public properties such as roads, highways, railroads, electric transmission lines, and gas pipelines.

SBC/SBN   Subscriber Based Criterion/ Subscriber Based Norms—Method of allocating additional spectrum to mobile operators based on subscriber numbers.

SDCA   Short Distance Charging Area—The area that covers one or more taluka or other administrative areas, normally extending from 800—2,000 sq kms, coinciding with Number Plan Area. Each SDCA is assigned a unique 3/4-digit trunk code. Wire-line subscribers have to dial only the subscribe numbers for communicating with other wire-line terminals within SDCAs.

SDH   Synchronous Digital Hierarchy—World standard designed for synchronous optical networks, speed of which ranges from 155 Mbps (often referred to as Synchronous Transport Mode STM-1)

SIM   Subscriber Identification Module—It is an integrated circuit that securely stores the International Mobile Subscriber Identity (IMSI) used to identify a subscriber on mobile telephony devices. A SIM is held on a removable SIM card often provided by the mobile operator, which can be transferred between different mobile devices. A SIM card contains its unique serial number known as Integrated Circuit Card Identifier (ICCID), security authentication and ciphering information, temporary information related to the local network, a list of the services the user has access to and passwords.

SIP	Session Initiation Protocol—A signalling protocol used to establish sessions over IP networks, such as those for telephone calls, audio conferencing, instant messaging between devices.
Smartphone	Mobile handset that has features such as QWERT key board, touch screen, FM radio, NFC, Internet browser, Wi-Fi, 3G/4G and GPS network connectivity. These handsets have powerful processors that mimic almost a PC and hence are able to provide rich user experience. The handset may use a proprietary operating system such as iOS as in Apple iPhones, Symbian as in Nokia phones, or open source operating systems such as Google's Android.
SMB	Small and Medium size Business—Businesses that normally have about 1,000 employees and less than Rs 500 crores turnover.
SMS	Short Message Service—Short, 160-character text messages that can be transmitted over digital cellular networks.
SoA	Service Oriented Architecture—A set of flexible design principles for integrating applications and services across different business domains and providing interoperability.
SOFA	State Optic Fibre Agency—A state agency jointly held by NOFA and respective State Government to plan, install, operate and maintain access aggregation network within the state to connect various access networks.
SSA	Secondary Service Area—The area divisioned by the Telegraph Authority and is co-terminus with the LDCA.
STB	Set Top Box—The device for receiving digital content over DTH or Cable TV with authorized access to subscribed channels. Authorization is enabled through smart cards given by the service provider.
TAMA	Time Assigned Multiple Access—The scheme used in satellite channels allocation using a pre defined time slots for each user terminal, either randomly or round-robin fashion.
TCP	Transmission Control Protocol—Developed by the University of California, Berkeley, USA as a transport layer protocol for the Internet. The protocol supports synchronous and continuous communication of packets between two end points on the network as opposed to

the alternative User Datagram Protocol that supports asynchronous and discontinuous transmission of packets. TCP is implemented at the end nodes of the network and provides error and flow control.

TDMA   Time Division Multiple Access—Process of allowing mobile radios to share radio frequency allocation by dividing up that time slots that are shared between simultaneous users of the radio channel. When a mobile radio communicates with a TDMA system, it is assigned a specific time position on the radio channel.

TDSAT   Telecom Dispute Settlement Appellate Tribunal—A dispute settlement body set up within TRAI to adjudicate disputes between different entities in the telecom sector.

TD-SCDMA Time Division—Synchronous Code Division Multiple Access—High speed 3G mobile communication standard that combines support for both circuit-switched and packet switched data. The standard combines Time Division Multiple Access (TDMA) with an adaptive, synchronous-mode Code Division Multiple Access (CDMA) component. TD-SCDMA was developed by the China Academy of Telecommunications Technology (CATT) in collaboration with Datang and Siemens.

TRAI   Telecom Regulatory Authority of India—The regulatory organization created on 28 March 1997 under 'The Telecom Regulatory Authority of India Act 1997' to regulate telecommunication services in the country. More information can be found at www.trai.gov.in

UASL   Unified Access Service Licence—Type of access service licence formulated in 2003 by the Government of India to provide any service using any technology.

UASP   Unified Access Service Provider—Telecom operator who has UASL.

UHF   Ultra High Frequency—Designates the ITU radio frequency range of electromagnetic waves between 300 MHz and 3 GHz (3,000 MHz), also known as the decimetre band or decimetre wave as the wavelengths range from one to ten decimetres (10 cm to 1 metre).

UMTS   Universal Mobile Telecommunications System—Set of wireless communication standards that enable the use of a combination of wireless and fixed systems in an effort to provide seamless telecommunications services to its users. The specifications for the UMTS system are overseen by the 3GPP.

USO   Universal Service Obligation—The telecom policy that enables the availability of telecommunication services in a non-discriminatory manner to all citizens of the country. Normally funded by USO Fund, collected through USO Levy (UASL) collected from the telecom operators.

VAS   Value-added Service—Services that provides benefits to a customer that are not part of the standard telecommunications services provided by the telecom operators.

VHF   Very High Frequency—The radio frequency range from 30 MHz to 300 MHz traditionally used for FM radio broadcasting.

VoIP   Voice over Internet Protocol—Process of sending voice traffic in the form of packets over IP networks. VoIP digitizes analog voice, compresses it, packetizes and sends it to the receiver.

VPN   Virtual Private Network—A network that provides the functions and features of a private network such as reliability and security without the need for dedicated private leased lines across communication sites. The communication is overlaid on public networks such as PSTN/ Internet. The service is typically provided by the ISPs.

VSAT   Very Small Aperture Terminal—These are geo-stationary satellite based user terminals consisting of the antennae and in-door electronics normally mounted on roof tops to cater to the data communication needs of dispersed users. The antennae are normally large enough to capture radio frequency beams in the Extended C band for TV broadcasting and Ku band for data communications and DTH.

VSNL   Videsh Sanchar Nigam Limited—Formed in 1986 by the Government of India to provide ILD and Internet services, later acquired by Tata Group and christened as Tata Communications.

WAN	Wide Area Network—A communication network that spans across distant geographical locations. WAN normally connects many different LANs through wide area transmission links.
WCDMA	Wideband Code Division Multiple Access—Wideband code division multiple access is a 3rd generation mobile communication system that uses Code Division Multiple Access (CDMA) technology over a wide frequency band to provide high-speed multimedia and efficient voice services. The WCDMA infrastructure is compatible with GSM mobile radio communication system.
Wi-Fi	Wireless Fidelity—Local area wireless networks based on IEEE 802.11 b/g/n standards operating in the licence free Industrial Medical and Scientific (ISM) radio frequency bands of 2.4 GHz and 5 GHz. Internet services using this technology is normally provided by the ISPs.
WiMAX	Worldwide interoperability for Microwave Access—Technology and protocols envisioned and supported by WiMAX forum, later adopted by IEEE as 802.16 standard for high-speed fixed and mobile wireless service, typically in spectrum blocks in 2 GHz to 11 GHz range. More information can be found at www.wimaxforum.org
WIPO	World Intellectual Property Organization is a specialized agency of the United Nations. WIPO was established by the WIPO Convention in 1967 with a mandate from its Member States to promote the protection of IP throughout the world through cooperation among states and in collaboration with other international organizations. Its headquarters are in Geneva, Switzerland. WIPO settles international domain name disputes. More information can be found at www.wipo.int.
WLL	Wireless Local Loop—Telecommunication access service provided through licensed Radio Frequencies, normally intended to be fixed at the customer premise, traditionally provided by BTOs.
WLL-LM	Wireless Local Loop—Limited Mobility—The WLL services that is limited within SDCAs.
WPC	Wireless Planning and Coordination wing—The unit in the DoT responsible for radio frequency spectrum planning and allocation.

References

Chapter 1

Economides, N. 1996. 'The Economics of Networks', *International Journal of Industrial Organization*, 14: 673–99.

Gurbaxani, V. 1990. 'Diffusion in Computing Networks: The Case of BITNET', *Communications of the ACM*, 33: 65–75.

InfoDev. 2000. *Telecommunications Regulation Handbook*. World Bank, Washington DC.

Kathuria, R., M. Uppal, and Mamta. 2009. 'An Econometric Analysis of the Impact of Mobile', Vodafone Policy Paper Series, 9: 5–20.

Katz, Michael L. and Carl Shapiro. 1985. 'Network Externalities, Competition, and Compatibility', *American Economic Review*, 75: 424–40.

Nuechterlein, J. and P. Weiser. 2005. *Digital Crossroads*. Cambridge, MA, USA: MIT Press.

Prasad, R. and V. Sridhar. 2008. 'Optimal Number of Mobile Service Providers in India: Tradeoff between Efficiency and Competition', *International Journal of Business Data Communications and Networking*, 4(3): 69–81.

Prasad, R. and V. Sridhar. 2009. 'Allocative Efficiency of the Mobile Industry in India and its implications for Spectrum Policy', *Telecommunications Policy*, DOI: 10.1016/j.telpol.2009.06.001.

Rai, A., T. Ravichandran, and S. Samaddar. 1998. 'How to Anticipate the Internets' Global Diffusion', *Communications of the ACM*, 41(10): 97–106.

Reed, D. 1999. That Sneaky Exponential—Beyond Metcalfe's Law to the Power of Community Building. Available at: http://www.reed.com/ accessed on 20 June 2009.

Reed, D. Feb 2000. 'The Law of the Pack', *Harvard Business Review*, 23–4.

Rogers, E. 1983. *Diffusion of Innovations*. New York: The Free Press.

Rouvinen, P. 2006. 'Diffusion of Digital Mobile Telephony: Are Developing Countries Different?', *Telecommunications Policy*, 30: 46–63.

Singh, S.K. 2008. 'The Diffusion of Mobile Phones in India', *Telecommunications Policy*, 32(9–10): 642–51.

Sridhar, Kala and V. Sridhar. 2006. 'Telecommunications and Growth: Causal Model, Quantitative and Qualitative Evidence', *Economic and Political Weekly*, 24 June, pp. 2611–19.

Sridhar, V. 2004. 'Network Externalities in Telecom', *Economic and Political Weekly*, 7 August.

Sridhar, V. 2005. 'Can We beat the S-curve Syndrome?', *The Financial Express*, 2 August.

Sridhar, V. 2005. 'Empower the regulator', *Economic Times*, 22 September.

Sridhar, V. 2005. 'Saturation Threat in Mobile Services', *Business Standard*, 31 October.

Sridhar, V. 2009. 'An Econometric Analysis of Mobile Services Growth Across Regions of India', *Netnomics*, DOI: 10.1007/s11066-009-9041-6.

Sridhar, V. and A. Malik. 2007. 'Turning Copper into Gold: Bharti Airtel's Fixed Line Service in India', Asia Case Research Centre, University of Hong Kong, Case Reference No: 07/329C. Also available through Harvard Business Publishing (Product Number: HKU 674).

Sridhar, V. April 2007. 'Growth of Mobile Services Across Regions of India', *Journal of Scientific and Industrial Research*, 66: 281–9.

Sridhar, V. and S.R. Raja. 2008. 'Move Over to Foundations and Alliances', *Economic Times*, 10 September.

Sridhar, V. and G. Venkatesh. 2010. 'Gravy Train Stops at App Platform', *Economic Times*, 1 March.

Sridhar, V. and S.R. Raja. 2008. 'Move Over to Foundations and Alliances', *Economic Times*. Accessed at: http://www.econmictimes.com/ on 10 September 2008.

Steinbock, D. 2001. 'Assessing Finland's Wireless Valley: Can the Pioneering Continue?', *Telecommunications Policy*, 25: 71–100.

Telecom Regulatory Authority of India (TRAI). 2007. 'A Journey Towards Excellence in Telecommunications'. Available at: http://www.trai.gov.in accessed on 10 January 2008.

The Wall Street Journal (WSJ). 1999. 'Iridium's Downfall: The Marketing Took A Back Seat to Science', 18 August.

Wang, M. and W. Kettinger. 1995. 'Projecting the Growth of Cellular Communications', *Communications of the ACM*, 38: 119–22.

Working Group on Telecom Sector for the Eleventh Five-Year Plan (WGTCOM). 2006. *Report of the Working Group on the Telecom*

Sector for the Eleventh Five-Year Plan (2007-2012), Department of Telecommunications, Ministry of Communications and IT, Government of India.

Yang, Y. 1997. *'Essays on Network Effects'*, PhD thesis, University of Utah, USA.

Chapter 2

Athreya, M.B. 1996. 'India's Telecommunications Policy', *Telecommunications Policy*, 20(1): 11–22.

Chowdary, T.H. 1998. 'Politics and Economics of Telecom Liberalization of India', *Telecommunications Policy*, 22(1): 9–22.

Department of Telecommunications (DoT). 2001. Guidelines for Issue of License for Basic Service. Retrieved from http://www.dotindia.com on 28 September 2003.

Department of Telecommunications (DoT). 2005. Clarification Regarding Fixed Wireless Terminal in UAS/Basic Service License. Accessed from www.dotindia.com on 31 October 2007.

Department of Telecommunications (DoT). 2006. 'Report of the Working Group on the Telecom Sector for the Eleventh Five Year Plan (2007–12). Accessed from www.dotindia.com on 31 October 2006.

Desai, A. 2006. Wireless Network Evolution: 2G to 3G. Prentice Hall.

Gupta, N. 2000. *The Business of Telecommunications*. New Delhi, India: Tata McGraw-Hill.

Hudson, H. 1994. 'Universal Service in the Information Age', *Telecommunications Policy*, 18(8): 658–67.

InfoDev. 2000. *Telecommunications Regulation Handbook*. World Bank, Washington DC.

InfoDev. Telecommunications Regulation Handbook. Retrieved from http://www.infodev.org/projects/314regulationhandbook/index.htm on 3 March 2003.

Investment Information and Credit Rating Agency (ICRA). 2002. 'Industrial Watch Series: The Indian Telecommunication Industry', Mumbai, India: ICRA.

International Telecommunications Union (ITU). 2003. *Yearbook of Statistics*. Geneva, Switzerland: ITU.

Jain, P. 2004. 'Techno-Economic Modeling of Basic Telecommunication Services—A System Dynamics Approach', FPM thesis, Indian Institute of Management, Lucknow, India.

Jain, R. 2001. 'A Review of Indian Telecom Sector', *India Infrastructure Report*, 189–235.

Jhunjhunwala, A. 2000. 'Unleashing Telecom and Internet in India', India Telecom Conference, Asia/Pacific Research Center, Stanford

University. Retrieved from http://www.tenet.res.in/papers/unleash. html on 3 March 2003.

Lechleider, J.W. 1991. 'Coordinated Transmiussion for Two-pair Digital Subscriber Lines', *IEEE Journal on Selected Areas in Communication*, 9(6): 920–30.

Melody, W.H. 2000. 'Shaping Liberalized Telecom Markets', *Telecommunications Policy*, 24: 803–6.

Mukherji, R. 2004. 'Privatization, Federalism and Governance', *Economic and Political Weekly*, 39(1): 109–113.

Mukherji, R. 2008. 'The Politics of Telecommunications Regulations: State–Industry Alliance Favouring Foreign Investment in India', *Journal of Development Studies*, 44(10): 1405–23.

Mukherji, R. 2010. 'The Political Economy of Development in India', in S. Ganguly and R. Mukherji (eds), *India Since 1980s*. New York: Cambridge University Press.

Petrazzini, B. 1996. 'Telecommunications Policy in India: The Political Underpinnings of Reform', *Telecommunications Policy*, 20(1): 39–51.

Sinha, N. 1996. 'The Political Economy of India's Telecommunications Reforms', *Telecommunications Policy*, 20(1): 23—38.

Sridhar, V. 2001. 'Finding a Way out of the Limited Mobility Imbroglio', *Economic Times*, 4 January.

Sridhar, V. 2003. 'Is it the Death Knell for Landlines?', *Economic Times*, 15 November.

Sridhar, V. 2003. 'Don't Blame The Regulator', *Economic Times*, 24 September.

Sridhar, V. 2004. 'Local Loop Unbundling: A New Dimension in Telecom Competition', *The Hindu Business Line*, 11 June.

Sridhar, V. 2005. 'Empower the Regulator', *Economic Times* 22 September.

Sridhar, V. 2006. 'Interconnect before One India', *Business Standard*. 27 March.

Sridhar, V. 2006. 'Should Local Loops be Unbundled', *Economic Times*. 17 July.

Sridhar, V. and A. Malik. 2007. 'Turning Copper into Gold: Bharti Airtel's Fixed Line Service in India', Asia Case Research Centre, University of Hong Kong, Case Reference No: 07/329C. Also available through Harvard Business Publishing (Product Number: HKU 674).

Telecom Regulatory Authority of India (TRAI). 2000. TRAI Amendment Ordinance. Retrieved from http://www.trai.gov.in on 1 November 2000.

Telecom Regulatory Authority of India (TRAI). 2003. The Telecommunication Interconnection Usage Charges Regulation. Retrieved from http://www.trai.gov.in on 29 April.

Telecom Regulatory Authority of India (TRAI). April 2004. Broadband India: Recommendations on Accelerating Growth of Internet and

Broadband Penetration. Available at http://www. trai. gov. in, accessed on 10 May 2004.

Telecom Regulatory Authority of India (TRAI). June 2005. Consultation Paper on Allocation and Pricing of Spectrum for 3G Services and Broadband Wireless Access. Available at http://www. trai. gov. in, accessed on 10 July 2005.

Telecom Regulatory Authority of India (TRAI). 2005. Recommendations on the Growth of Telecom Services in Rural India. Accessed http://www.trai.gov.in on 29 November 2005.

Telecom Regulatory Authority of India (TRAI). 2006. TRAI Issues Amendment to Interconnect Usage Charges Regulation. Retrieved from http://www.trai.gov.in on 29 April 2006.

Telecom Regulatory Authority of India (TRAI). 2007. Draft Recommendations on Growth of Broadband. Retrieved from http://www.trai.gov.in on 20 September 2007.

Universal Service Obliagtion Fund (USO). 2011. *The Year-wise Allocation and Distribution of USO Fund.* Accessed http://www.dot.gov.in/uso/implementationstatus.htm on 8 June 2011.

Waring, D.L., J.W. Lechleider, and T.R. Hsing. 1991. 'Digital Subscriber Line Technology Facilitates a Graceful Transition from Copper to Fiber', *IEEE Communications Magazine*, 29(3): 96-104.

Chapter 3

Buchanan, K., R. Fudge, D. McFarlene, T. Phillips, A. Sasaki, and H. Xia. 1997. 'IMT-2000: Service Provider's Perspective', *IEEE Communication Magazine*, 8–13.

Carsello, R. 1997. 'IMT-2000 Standards: Radio Aspects', *IEEE Personal Communications*, 30–40.

CRISINFAC. 2006. *Telecom Services Annual Review.*

Dekleva, S., J.P. Shim, U. Varshney, and G. Knoerzer. 2007. 'Evolution and Emerging Issues in Mobile Wireless Networks', *Communications of the ACM*, 50(6): 38–43.

Department of Telecommunications (DoT). December 2005. Guidelines for Unified Access Service License. Retrieved from http://www.dot.giv.in on 1 November 2008.

Department of Telecommunications (DoT). 2008a. Guidelines for Auction and Allotment of Spectrum for 3G Telecom Services. Retrieved from http://www.dot.giv.in on 1 November 2008.

Department of Telecommunications (DoT). 2008b. Detailed Guidelines for Auction and Allotment of Spectrum for BWA Services. Retrieved from http://www.dot.giv.in on 1 November 2008.

Department of Telecommunications (DoT). October 2009. Auction of 3G and BWA Spectrum: Auction Rules. Retrieved from http://www.dot.gov.in on 11 June 2010.

Desai, A. 2006. *India's Telecommunications Industry: History, Analysis, Diagnosis.* New Delhi: Sage Publications.

Garg, V. 2001. *Wireless Network Evolution: 2G to 3G.* Prentice Hall.

International Telecommunications Union (ITU). 2003. *Yearbook of Statistics: Telecommunication Services Chronological Time Series 1992-2001.* Geneva, Switzerland: ITU.

Investment Credit Rating Agency (ICRA). June 2002. *Industrial Watch Series: The Indian Telecommunication Industry.* ICRA: Mumbai, India.

Jain, R.S. 2001. 'Spectrum Auctions in India: Lessons from Experience', *Telecommunications Policy,* 25: 671–88.

Jain, P. and V. Sridhar. 2003. 'Analysis of Competition and Market Structure of Basic Telecommunication Services in India', *Communications and Strategies,* 52: 271–93.

Loo, B.P.Y. 2004. 'Telecommunications Reforms in China: Towards an Analytical Framework', *Telecommunications Policy,* 28: 697–714.

Marbridge. 2011. *Mainland China Telecom Statistics.* Retrieved from http://www.marbridgeconsulting.com/statistics.html on 1 March 2011.

Mathew, L., J. Mysore, and V. Nair. 2006. 'Is the Indian Market Ready to go "Virtual" for Mobile?', *Diamond Management and Technology Consultants Report.*

McDowell, S. and J Lee. 2003. 'India's Experiments in Mobile Licensing', *Telecommunications Policy,* 27: 371—82.

Mukerji, R. 2008. 'The Politics of Telecommunications Regulations: State–Industry Alliance, Favouring Foreign Investment in India', *Journal of Development Studies,* 44(10): 1405–23.

Pagare, P. and V. Sridhar. 2010. 'India Should Leapfrog to 4G', *Economic Times,* 17 March.

Paul Budde Communications Pvt. Ltd. (PBC). 2006. 'China—Mobile Market—Overview and Statistics'.

Prasad, R. and V. Sridhar. 2008. 'Optimal Number of Mobile Service Providers in India: Trade-off between Efficiency and Competition', *International Journal of Business Data Communications and Networking,* 4(3): 69–81.

Prasad, R. and V. Sridhar. 2009. '3G Delay Hurting the Economy', *Economic Times,* 9 February.

Rapport, T. 1996. *Wireless Communications Principles and Practice.* New Jersey, USA: Prentice-Hall, Inc.

Ratnoo, R. and V. Sridhar. 2010. 'Mobiles for Delivery of Public Services', *Economic Times,* 29 April.

Singh, H. 'Motorola Kicks off Price War in Handset', *Financial Express*, 28 November.

Singh, S.K. 2008. 'The Diffusion of Mobile Phones in India', *Telecommunications Policy*. doi:10.1016/j.telpol.2008.07.005: 1–10.

Sridhar, V. 2003. 'Tariff Forbearance for Competition', *Economic Times*, 19 May.

Sridhar, V. 2006. 'Government Operators should be Hauled up as Well', *The Financial Express*, 13 March.

Sridhar, V. 2006. 'Interconnect before One India', *Business Standard*, 27 March.

Sridhar, V. and S.R. Raja. 2008. 'The Next Big Mobile Wave', *Economic Times*, 16 July.

Sridhar, V. and S.R. Raja. 2010. 'What's Missing?', *Business Line*, 4 January.

Sridhar, V. and Kala Sridhar. 2010. 'One Paisa Revolution', *Financial Express*, 4 August.

Stallings, W. 2004. *Business Data Communications*. New Delhi: Pearson Education.

Telecom Regulatory Authority India (TRAI). 1999. Consultation Paper (99/6) on Issues Relating to Cellular Mobile Services. Retrieved from http://www.trai.gov.in on 21 November 2006.

Telecom Regulatory Authority India (TRAI). 2001. Survey Conducted by TRAI to Assess Quality of Service Provided by BSOs and CMSOs. Retrieved from http://www.trai.gov.in on 15 December 2001.

Telecom Regulatory Authority India (TRAI). 2003. Telecommunications Interconnect User Charges Regulation. Retrieved from http://www.trai.gov.in on 30 October 2003.

Telecom Regulatory Authority of India (TRAI). 2005a. Recommendations on Spectrum Related Issues. Retrieved from http://www.trai.gov.in on 13 May 2005.

Telecom Regulatory Authority of India (TRAI). 2005b. Study Paper on Financial Analysis of Telecom Industry of China and India. Retrieved from http://www.trai.gov.in on 16 July 2005.

Telecom Regulatory Authority of India (TRAI). 2005c. Recommendations on the Growth of Telecom Services in Rural India. Retrieved from http://www.trai.gov.in on 3 October 2005.

Telecom Regulatory Authority of India (TRAI). 2005d. Study Paper (No. 212005) on Indicator for Telecom Growth. Retrieved from http://www.trai.gov.in on 3 October 2005.

Telecom Regulatory Authority of India (TRAI). 2005e. The Indian Telecom Services: Performance Indicators October–December 2005.

Telecom Regulatory Authority of India (TRAI). March 2006. Recommendations on Mobile Number Portability. Retrieved from http://www.trai.gov.in on 1 May 2006.

Telecom Regulatory Authority of India (TRAI). 2006a. Amendments to Interconnect Usage Charges. Retrieved from http://www.trai.gov.in on 15 February 2006.

Telecom Regulatory Authority of India (TRAI). 2006b. Consultation Paper (912006) on Allocation and Pricing of Spectrum for 3G Services and Broadband Access. Retrieved from http://www.trai.gov.in on 11 June 2006.

Telecom Regulatory Authority of India (TRAI). 2006c. The Indian Telecom Services Performance Indicators for the Financial Year ending 31 March 2006. Retrieved from http://www.trai.gov.in on 7 September 2006.

Telecom Regulatory Authority of India (TRAI). June 2006. Allocation and Pricing of Spectrum for 3G and Broadband Wireless Access Services. Retrieved from http://www.trai.gov.in on 30 November 2006.

Telecom Regulatory Authority of India (TRAI). January 2009. Draft Recommendations on the Growth of Value-Added Services and Regulatory Issues. Retrieved from http://www.trai.gov.in on 1 February 2009.

Universal Service Obligation Fund (USO). 2011. *The Year-wise Allocation and Distribution of USO Fund*. Accessed http://www.dot.gov.in/uso/implementationstatus.htm on 8 June 2011.

Vriendt, J. *et al*. 2002. 'Mobile Network Evolution: A Revolution on the Move', *IEEE Communication Magazine*, 104–111.

Whalley, J. and P. Curwan. 2006. 'Third-generation New Entrants in the European Mobile Telecommunications Industry', *Telecommunications Policy*, 30: 622–32.

Yuan, X., W. Zheng, Wang, Y., Z. Xu, Q. Yang, and Y. Gao. 2006. 'Xiaolingtong versus 3G in China: Which will be the Winner?', *Telecommunications Policy*, 30: 297–313.

Chapter 4

Bykowski, M. 2003. 'A Secondary Market for the Trading of Spectrum: Promoting Market Liquidity', *Telecommunications Policy*, 27: 533–41.

Baumol, W.J. 1982. 'Contestable Markets: An Uprising in the Theory of Industry Structure', *American Economic Review*, 72(1): 1–15.

Chapin, J.M., and W.H. Lehr. 2007. Cognitive Radios for Dynamic Spectrum Access—The Path to Market Success for Dynamic Spectrum Access Technology. *IEEE Communications Magazine*, 45(5): 96–103.

Crocioni, P. 2009. 'Is Allowing Trading Enough? Making Secondary Market in Spectrum Work', *Telecommunications Policy*, 33(8): 451–68.

Department of Telecommunications (DoT). 2003. Guidelines for Unified Access Service License. Retrieved from http://www.dotindia.com on 15 January 2006.

Department of Telecommunications (DoT). 2005. Guidelines for Unified Access Service License. Retrieved from http://www.dotindia.com on 15 January 2006.

Department of Telecommunications (DoT). March 2008. Guidelines for Infrastructure Sharing. Retrieved from http://www.dotindia.com.

Department of Telecommunications (DoT). August 2008. Guidelines for Auction and Allotment of Spectrum for 3G Telecom Services. Retrieved from http://www.dotindia.com on 10 August 2008.

Falch, M. and R. Tadayoni. 2004. 'Economic Versus Technical Approaches to Frequency Management', *Telecommunications Policy*, 28: 197–211.

Gruber, H. 2007. 3G Mobile Telecommunication Licenses in Europe: A Critical Review. DOI: 10.1108/146366907108276777.

Intven, H., J. Oliver and E. Sepulveda. 2000. Telecommunications Regulation Handbook. Retrieved 15 Jan 2001 from http://www.infodev.org/projects/314regulationhandbook.

Jain, R. and R. Jain. 2007. Spectrum Refarming: Digital Switchover of US TV Broadcast Signals. Retrieved from http:// www.iimahd.ernet.in/iitcoe/ on 25 December 2007.

Prasad, R. and V. Sridhar. 2007. 'Spectrum Allocation Mechanism for 3G Mobile Services', *Economic and Political Weekly*, 9 June, 42(23).

Prasad, R. and V. Sridhar. 2008. 'Spectrum Policy too Narrowly Focussed', *Economic Times*, 1 June.

Prasad, R. and V. Sridhar. 2008. 'A Critique of Spectrum Management in India', *Economic and Political Weekly*, 20 September, 13–17.

Prasad, R. and V. Sridhar. 2008. 'Optimal Number of Mobile Service Providers in India: Trade-off between Efficiency and Competition', *International Journal of Business Data Communications and Networking*, 4(3): 69–81.

Prasad, R. and V. Sridhar. 2008. 'Are Operators Hoarding Spectrum', *Economic Times*, 19 May.

Sridhar, V. 2006. 'To Allocate Spectrum, Study Real Estate', *Financial Express*, 29 September.

Sridhar, V. 2007. 'Growth of Mobile Services across Regions of India', *Journal of Scientific and Industrial Research*, 66: 281–9.

Sridhar, V. and R. Prasad. 2011. 'Towards a New Policy Framework for Spectrum Management in India', *Telecommunications Policy*, 35: 172–84. DOI: 10.1016/j.telpol.2010.12.004.

Telecom Regulatory Authority of India (TRAI). 2005. Recommendations on Spectrum-Related Issues. Retrieved from http://www.trai.gov.in on 13 May 2005.

Telecom Regulatory Authority of India (TRAI). 2006. Recommendations on Spectrum Allocation and Pricing for 3G and BWA Services. Retrieved from http://www.trai.gov.in on 13 May 2007.

Telecom Regulatory Authority of India (TRAI). 2007. Recommendations on Reviewing of License Terms and Conditions and Capping of Number of Access Providers. Retrieved from http://www.trai.gov.in on 13 November 2007.

Valletti, T.M. 2001. 'Policy Forum: Spectrum Trading', *Telecommunications Policy*, 25: 655–70.

Wireless Planning and Coordination (WPC). 2008. Order No J-4025/200(17)/2004-NT dated 17 January 2008. Retrieved from World Wide Web: http://www.wpc.dot.gov.in on 9 May 2008.

Xavier, P. and D. Ypsilanti. 2006. 'Policy Issues in Spectrum Trading', *Info*, 8(2): 34-61, doi: 10.1108/14636690610653581.

Chapter 5

Business Standard. 2008. 'Adag's Flag Wins Appeal Against VSNL: Hague Court Upholds 2006 Ruling on Cable Upgrade Dispute', available at: http://www.businessstandard.com, accessed on 3 December 2008.

Department of Telecommunications (DoT). 2001. Guidelines for Issue of License for National Long-Distance Service. Available at http://www.dotindia.in, accessed on 20 September 2008.

Department of Telecommunications (DoT). 2002. Guidelines for Issue of License for International National Long-Distance Service No.10-19/2001-BS-I. Available at http://www.dotindia.in, accessed on 20 September 2008.

Department of Telecommunications (DoT). 2003. National Numbering Plan. Available at http://www.dotindia.in, accessed on 20 September 2008.

Department of Telecommunications (DoT). December 2005. Guidelines for Issue of License for International Long-Distance Service. Available at http://www.dotindia.in, accessed on 20 August 2007.

Department of Telecommunications (DoT). July 2009a. List of ILD Licensees. Available at http://www.dotindia.in, accessed on 28 February 2011.

Department of Telecommunications (DoT). July 2009b. List of ILD Licensees. Available at http://www.dotindia.in, accessed on 28 February 2011.

Esselaar, S., A. Gillwald, and E. Sutherland. 2007. The Regulation of Undersea Cables and Landing Stations. Available at: http://link.wits.ac.za, accessed on 10 December 2008.

Federal Communications Commission (FCC). 2002. US IMTS Net Settlement payment (1985–2000). Available at http://www.fcc.gov/ib/pd/pf/nsp.xls accessed on 19 November 2008.

Federal Communications Commission (FCC). 2006. International Settlements Policy and US-International Accounting Rates. Available at: www. fcc.gov/ib/pd/pf/account.html, accessed on 18 November 2008.

Federal Communications Commission (FCC). 2008. International Settlements Policy and US International Accounting Rates. Available at www.fcc.giv/ib/pd/pf/account.html., accessed on 18 November 2008.

Shy, Oz. 2001. *The Economics of Network Industries*. Cambridge, UK: Cambridge University Press.

Sridhar, Kala and V. Sridhar. 2005. 'The Substitution Effect on Access Deficit Charge', *Business Line*, 1 December.

Sridhar, V. 2000. 'How Much Should a Call to the US Cost?', *Economic Times*, 8 September.

Sridhar, V. 2001. 'VSNL: A Golden Goose or a Lame Duck?', *Economic Times*, 17 April.

Sridhar, V. 2002. 'Calling New York Cheaper than Dialling Delhi?', *The Hindu Business Line*, 13 September.

Sridhar, V. 2004. 'ILD Turning Grey', *The Financial Express*, 14 October.

Sridhar, V. 2005. 'More May not be Merrier...', *The Financial Express*, 11 October.

Sridhar, V. 2005. 'How Should NLD be Opened Up? Implement Unified Licensing Regime', *Economic Times*, 17 October.

Sridhar, V. 2006. 'Furthering Competition on Long-distance Service', *The Financial Express*, 28 June.

Sridhar, V. 12 September 2006. 'Mega Telecom Partnerships', *The Financial Express*. Available at: www.financialexpress.com, accessed on 12 September 2006.

Telecom Regulatory Authority of India (TRAI). December 2001. Issues Relating to Interconnection between Access Providers and National Long-Distance Operators. Available at http://www.trai.giv.in, accessed on 10 January 2008.

Telecom Regulatory Authority of India (TRAI). October 2003. 'The Telecommunication Interconnection Usage Charges Regulation.' Available at: http://www.trai.gov.in, accessed on 20 August 2006.

Telecom Regulatory Authority of India (TRAI). September 2005. 'Telecommunications Tariff Order Amendment No.310-3(1)/2003-Eco'. Available at http://www.trai.giv.in, accessed on 25 November 2008.

Telecom Regulatory Authority of India (TRAI). December 2005. Recommendations on Measures to Promote Competition in International Private Leased Circuits in India. Available at http://www.trai.giv.in, accessed on 25 November 2008.

Telecom Regulatory Authority of India (TRAI). March 2007. Recommendations on Terms and Conditions for Resale in International Private Leased Circuits (IPLC) Segment. Available at http://www.trai.giv.in, accessed on 5 December 2008.

Telecom Regulatory Authority of India (TRAI). March 2008. Recommendation on Support for Rural Wireline Connections Installed before 1.4.2002 from USOF and Phasing out of ADC. Available at http://www.trai.giv.in, accessed on 25 November 2008.

Telecom Regulatory Authority of India (TRAI). May 2008. Consultation Paper on Carrier Selection. Available at: http://www.trai.giv.in.

Telecom Regulatory Authority of India (TRAI). August 2008. Recommendations on Provision of Calling Cards by Long-distance Operators. Available at: http://www.trai.gov.in.

Telecom Regulatory Authority of India (TRAI). October 2008. The Indian Telecom Service Performance Indicators: April–June, 2008. Available at http://www.trai.gov.in, accessed on 10 November 2008.

Varghese, S. 2006. Reliance Infocomm's Strategy and its Impact on the Indian Mobile Telecommunications Scenario. Department of Media and Communications, London School of Economics.

Videsh Sanchar Nigam Limited (VSNL). 2005. Reply Comments of VSNL to Section 1377 Request. Available at http://www.tatacommunications. com, accessed on 3 December 2008.

Voice and Data. 2002. Videsh Sanchar Nigam Limited: Gritful Stances. Available at http://www.voicendata.com, accessed on 24 October 2008.

Voice and Data. 2003. Videsh Sanchar Nigam Limited: Gritful Stances. Available at http://www.voicendata.com, accessed on 24 October 2008.

Voice and Data. 2008. *Voice and Data Volume II 2008*. Available at http://www. voicendata.com, accessed on 24 October 2008.

Voice and Data. 2010a. India's NLD market has grown by 13.6% in FY 2009-10. Available at http://www.voicendata.com, accessed on 26 February 2010.

Voice and Data. July 2010b. ILD: Picking up pace. Available at http://www. voicendata.com, accessed on 26 February 2010.

Chapter 6

Cyber Law India. 2008. Some Cyber Law Perspectives. Accessed http:// www.cyberlaws.net on 14 January 2008.

Department of Telecommunications (DoT). 2005. Guidelines and General Information for Internet Service Providers (ISP) *(No.845-51/97-VAS)*. Retrieved from http://www.dotindia.com on 14 July 2007.

Department of Telecommunications (DoT). 2007. Guidelines and General Information for Internet Service Providers (ISP) *(No.820-*

1/2006-LR). Retrieved from http://www.dotindia.com on 30 August 2007.

Gregory, P. 2006. *SIP Communications for Dummies* (Avaya Custom Edition). Hoboken, NJ, USA: Wiley.

International Corporation of Assigned Names and Numbers (ICANN). 2008. *Internationalized Domain Names*. www document, http://idn. Icann.org, accessed 16 January 2008.

Internet World Stats. 2010. World Internet Usage and Population Statistics: Dec 2010. Available at http://www.internetworldstats.com/stats.htm, accessed on 15 Jan 2011.

Investment Information and Credit Rating Agency (ICRA). 2000. *Industry Watch Series: The Indian Internet Business Report*. New Delhi: ICRA Limited.

Jain, R., R. Kathuria, R. Dass, and S. Sinha. 2010. Research and Action Agenda for a National Broadband Initiative. Report by ICREAR and IIMA IDEA Telecom Centre of Excellence.

National Internet Exchange of India (NIXI). 2007. '.in Domain Name Registry', accessed http://www.nixi.in on 12 December 2007.

Sheldon, T. 2001. *Encyclopedia of Networking and Telecommunications*. Delhi: Tata McGrawHill.

Sridhar, V. 2003. 'The Next Generation of Internet Addresses', *Business Standard*, 22 October.

Sridhar, V. 2004. 'A New Dimension in Telecom Competition', *The Hindu Business Line*, 11 June.

Sridhar, V. 2005. 'Making VPN Services Cost-Effective for Users', *Business Line*, 19 August.

Sridhar, V. 2006. 'Don't Try Nailing Jelly to the Wall', *Financial Express*, 19 December.

Sridhar, V. 2009. 'Regulatory Dilemma Over VoIP', *Economic Times*, 4 March.

Sridhar, V. and A. Malik. 2007. Turning Copper into Gold: Bharti Airtel's Fixed Line Service in India. Asia Case Research Centre, University of Hong Kong, Case Reference No: 07/329C, Available at www.acrc. org.hk. Also available through Harvard Business Publishing (Product Number: HKU 674) at hbsp.harvard.edu

Stallings, W. 2001. *Business Data Communications* (Fourth Edition). Pearson Education.

Tanenbaum, A. 1996. *Computer Networks*. New Jersey, USA: Prentice Hall.

Telecom Regulatory Authority of India (TRAI). October 2003. *Telecommunications Internet User Charges Regulation*. www document, http://www. trai. gov. in, accessed on 10 December 2003.

Telecom Regulatory Authority of India (TRAI). April 2004. Broadband India: Recommendations on Accelerating Growth of Internet and

Broadband Penetration, available at http://www. trai. gov. in, accessed on 10 May 2004.

Telecom Regulatory Authority of India (TRAI). August 2005. Recommendations on Entry fee and License fee for ISP License with Virtual Private Network, available, at www document, http://www. trai. gov. in, accessed on 7 August 2005.

Telecom Regulatory Authority of India (TRAI). April 2007. Recommendations on Improvement in the Effectiveness of National Internet Exchange of India (NIXI), available at, http://www. trai. gov. in, accessed on 7 May 2007.

Telecom Regulatory Authority of India (TRAI). May 2007. Recommendations on Review of Internet Services. Retrieved on 7 August 2007 from http://www.trai.gov.in.

Telecom Regulatory Authority of India (TRAI). August 2008. Recommendations on Issues Related to Internet Telephony. Retrieved on 15 September 2008 from http://www.trai.gov.in.

Telecom Regulatory Authority of India (TRAI). December 2010. Recommendations on National Broadband Plan. Retrieved on 15 January 2011 from http://www.trai.gov.in.

Venkatesh, G. and V. Sridhar. 2009. 'Let the Traffic Flow'. *Business Line*, 14 September.

Voice and Data. July 2007. Internet and Broadband: Uphill Task. Available at http://www.voicedata.com.

Weiss, A. June 2006. 'Net Neutrality? There's Nothing Neutral about It', *netWorker*, 18–25.

World Intellectual Property Organization (WIPO). 2008. Cybersquatting Remains on the Rise with further Risk to Trademarks from New Registration Practices, available at, http://www.wipo.int/pressroom/en/articles/2007/article_0014.html, accessed 14 January 2008.

Chapter 7

Blèret, J., J-P. Dehaene, and P. Labaye. 1998. 'Aquila: A Satellite Access Network with Low Communication Costs', *Alcatel Communications Review*, 2nd Quarter, 124–130.

Chhibbar, N.K. 2000. *Asia Pacific Satellite—The Last Obstacle*. Available at: http://www.gvf.org/ solutions/articles/last_obstacle.doc, accessed on 1 April 2009.

Department of Telecommunications (DoT). 2007. Broad Guidelines for Issue of License for Commercial VSAT Service Providers and Captive VSAT Service. Available at: http://www.dotindia.in, accessed on 25 March 2009.

Indian Space Research Organization (ISRO). 2007. Available at: http://www.isro.org, accessed on 1 April 2007.

Gunter. 2007. Gunter's Space Page. Available at: http://space.skyrocket.de/, accessed on 1 April 2007.

Telecom Regulatory Authority of India (TRAI). October 2004. Recommendations on Issues Relating to Broadcasting and Distribution of TV Channels, Accessed from http://www.trai.gov.in on 19 November 2004.

Voice and Data. July 2001. VSAT: Policy Glitches. Available at: http://www.voicendata.com, accessed on 2 April 2009.

Voice and Data. 30 March 2003. VSATs: As You Like It. Available at: http://www.voicendata.com, accessed on 2 April 2009.

Voice and Data. July 2004. VSATs: Plain Vanilla Changes Flavors. Available at: http://www.voicendata.com, accessed on 2 April 2009.

Voice and Data. 13 June 2005. VSATs: Value Down; Volume Up. Available at: http://www.voicendata.com, accessed on 9 April 2009.

Voice and Data. 6 January 2005. VSATs: A Happy New Year. Available at: http://www.voicendata.com, accessed on 9 April 2009.

Voice and Data. March 2007. SBI: Today We Are Running a Network which is Probably Bigger than Some Network Service Providers. *V&D Gold Book.* Available at: http://www.voicendata.com, accessed on 2 April 2009.

Voice and Data. 5 March 2009. Let there be Satellite. Available at: http://www.voicendata.com.

Chapter 8

Contractor, N., A. Singhal, and E. Rogers. 1988. 'Metatheoretical Perspectives on Satellite Television and Development in India', *Journal of Broadcasting & Electronic Media*, 32(2): 129–48.

Department of Telecommunications (DoT). 2001. Guidelines for Obtaining License for Providing Direct-To-Home (DTH) Broadcasting Services in India. Accessed from http://www.dodindia.gov.in on 15 July 2006.

Department of Telecommunications (DoT). 2009. Guidelines for Providing Head-end In The Sky (HITS) Broadcasting Service in India. Accessed from http://www.dodindia.gov.in on 29 January 2011.

Digital TV (DTV). 2009. Digital Television Transition—12 June 2009. Retrieved from http://www.dtv.gov on 15 June 2009.

Sridhar, Kala and V. Sridhar. 2005. 'The Substitution Effect on Access Deficit Charge', *The Hindu Business Line*, 1 December.

Sridhar, Kala and V. Sridhar. 2008. The Politics and Economics of CAS, *The Business Standard*, 24 January.

Tanenbaum, A. 2001. *Computer Networks*. New Jersey, USA: Prentice Hall.

Telecom Regulatory Authority of India (TRAI). January 2004. Consultation Note on Issues Relating to Broadcasting and Cable Services. Retrieved on 10 August 2004 from http://www.trai.gov.in.

Telecom Regulatory Authority of India (TRAI). August 2004. Recommendations on Issues Relating to Broadcasting and Cable Services. Retrieved from http://www.trai.gov.in on 10 October 2004.

Telecom Regulatory Authority of India (TRAI). September 2005. Recommendations on Digitalization of Cable Television. Retrieved on 15 June 2009 from http://www.trai.gov.in.

Telecom Regulatory Authority of India (TRAI). October 2007. Recommendations on Head-end-In-The-Sky (HITS). Retrieved on 20 October 2007 from http://www.trai.gov.in.

Telecom Regulatory Authority of India (TRAI). January 2008a. Consultation Paper on Issues Relating to 3rd Phase of Private FM Radio Broadcasting. Retrieved on 15 January 2011 from http://www.trai.gov.in.

Telecom Regulatory Authority of India (TRAI). January 2008b. Recommendations on the Growth of Broadband. Retrieved on 10 January 2008 from http://www.trai.gov.in

Telecom Regulatory Authority of India (TRAI). 20 January 2008c. Recommendations on the Interoperability and Other Issues Relating to DTH. Retrieved on 26 May 2008 from http://www.trai.gov.in.

Telecom Regulatory Authority of India (TRAI). March 2008. Consultation Paper on Restructuring of Cable TV Services. Retrieved on 1 May 2008 from http://www.trai.gov.in.

Telecom Regulatory Authority of India (TRAI). July 2008. Recommendations on Restructuring of Cable TV Services. Retrieved on 10 August 2008 from http://www.trai.gov.in.

Telecom Regulatory Authority of India (TRAI). March 2009. Consultation Paper on DTH Issues relating to Tariff Regulation and New Issues under Reference. Retrieved on 10 March 2009 from http://www.trai.gov.in.

Chapter 9

Department of Telecommunications (DoT). 2004. Status Paper on Manufacture of Telecom Equipment in India. Retrieved from http://www.dot.gov.in/osp/Status%20paper%20on%20Manufacturing%20sector.doc.

Jhunjhunwala, A., B. Ramamurthy, and T. Gonsalves. 1998. 'The Role of Technology in Telecom Expansion in India', *IEEE Communications Magazine*, 36(11): 88–94.

Mani, S. 2005. Dragon vs. Elephant: A Comparative Analysis of Innovation Capability in the Telecommunications Equipment Industry in India and China. Retrieved from www.cds.in on 14 January 2011.

Martinez-Jerez, A. and V. G. Narayanan. 2006. 'Strategic Outsourcing at Bharti Airtel Limited. One Year Later', Case No: 9-107-003, Boston, MA: Harvard Business School.

Mody, R. and V. Sridhar. 10 January 2011. Changing World Order in IT, Telecom. Retrieved from www.economiestimes.com on 11 January 2011.

Singh, S. and R. Dubey. 2004. 'The World's Top Off-shoring Locations', *Business Standard*. Retrieved on 4 October 2004 from http://www.businessworldindia.com.

Sridhar, Kala and V. Sridhar. 2010. 'One-Paisa Revolution', *Financial Express*, 4 August.

Sridhar, V. 2009. 'Strategic Outsourcing: Opportunities and Challenges for Telecom Operators', in I. Bose (ed.). *Breakthrough Perspectives in Network and Data Communications Security, Design and Applications, Volume 1 of the Advances in Business Data Communications and Networking Series*. Hershey, PA, USA: IGI Global, 1-13. Also reprinted in K. Amant (ed.), *IT Outsourcing: Concepts, Methodologies, Tools, and Applications*, Hershey, PA, USA: IGI Global.

Telecom Regulatory Authority of India (TRAI). 2010. Consultation Paper on Encouraging Telecom Equipment Manufacturing in India. *Retrieved from* http:// www.trai.gov.in on 10 January 2011.

Voice and Data. 2007. C-DOT Alcatel: Realizing WiMax. Retrieved from http://voicendata.ciol.com/content/broadband/107020703.asp on 15 January 2011.

Voice and Data. November 2009. Destination R&D. Retrieved from http://voicendata.ciol.com/content/service_provider/109110205.asp on 19 January 2010.

Voice and Data. May 2010. R&D Efforts by Indian Handset Players. Retrieved from http://voicendata.ciol.com/content/Contributory Articles/110051001.asp on 19 January 2011.

Voice and Data. 2010. ITI: Surprise Package. Retrieved from http://www.voicendata.ciol.com/content/vnd100_2010vol-I/110060836.asp on 13 January 2011.

Chapter 10

Grover, V. and K. Saeed. 2003. 'The Telecommunication Industry Revisited', *Communications of the ACM*, 46(7): 119—125.

Prasad, R. and V. Sridhar. 2008. 'New Frontier of Telecom Policy', *Mint*, 14 August.

Sridhar, V. 2000. 'Convergence is the Key to Future Growth in Bracing for Tomorrow: A Response Feature', *Times of India*, 15 October.

Sridhar, V. 2003. 'Is Telecom Fit for Mega Mergers?', *Economic Times*, 22 December.

Sridhar, V. 2006. 'Mega Telecom Partnerships', *Financial Express*, 12 September.

Sridhar, V. 2007. 'Convergence: Evolution and Challenges', *Business Standard*, 20 March.

Sridhar, V. 2008. 'Blossoming Telecom Partnerships', *Economic Times*, 26 March.

Sridhar, V. 2009. 'Telecommunication Partnerships to Ponder', *Economic Times*, 9 September.

Sridhar, V. and G. Venkatesh. 2010. 'What's in store for Mobile Telecom?', *Business Line*, 28 June.

Telecommunications Regulatory Authority of India (TRAI). April 2011. Recommendations on Telecom Equipoment Manufacturing Policy. Available at http://www.trai.gov.in. Accessed on 15 June 2011

Tele.net. January 2010. Volume 11, Issue 1.

Voice and Data. June 2008. V&D, Issue 100, Volume 1.

Index